HORUS IN THE PYRAMID TEXTS

THOMAS GEORGE ALLEN

Atlas Occulta

ISBN: 978-1-955087-27-8

© 2022, Atlas Occulta

Al rights reserved. No part of this publication maybe reproduced, translate, store in a retrieval system, or transmitted in any form or by any means, electronic, mechanical, photocopying, recording or otherwise, without prior written permision from the publisher.

PROLOGUE

This work has been selected by scholars as being culturally important and is part of the knowledge base of civilization as we know it.

In a group of five pyramids at Sakkara, the tombs of Egypt's Fifth- and Sixth-Dynasty pharaohs, who ruled from about 2650 to 2475 B. C, is preserved as the largest body of inscriptions known anywhere from that remote age. These mortuary and religious texts not only present through their allusions an invaluable commentary on early Egyptian civilization, but they reveal the earliest discernible phases of man's thoughts on the great problems of cosmic origin and human destiny. Here lie the occurrences of divine names and epithets. Differentiation of independent deities from the mass proved a rather arbitrary matter. Not only do various types of supernatural beings appear, from the great. Cosmic powers down to representative of the animal and vegetable world geographic incarnations, and even inanimate objects. These texts shed light on 1.Terms regularly used as the principal designation of well-known deities. 2. Terms apparently used as principal designation of less-known deities, 3. Names of celestial bodies, animals, and serpents, appearing as supernatural agents, 4. Names of barques and crowns (unless supernatural serpent-element be involved). 5 An index of all occurrences in the Pyramid Texts of divine names. The most detailed myth traceable is that which records how Horus was born to Isis in Khemmis, fought with Set in his young manhood, and after recovering his eye? which Set. had taken and swallowed, bestowed it upon his father Osiris. The judicial proceedings that ensued at Heliopolis seem to have been a trial of Horus himself or, again, concerned the eye. They resulted in any case in the defeat of Set.

This scarce antiquarian book is a facsimile reprint of the original work. This title is an authentic reproduction of the original printed text in shades of gray and may contain minor errors. **IMPORTANT** Despite the fact that we have attempted to accurately maintain the integrity of the original work, the present reproduction may have minor errors beyond our control like: missing and blurred pages, poor pictures and markings. Because this book is culturally important, we have made available as part of our commitment to protect, preserve and promote knowledge in the world. This title is an authentic reproduction published by Thomas George Allen's 1916 Doctoral Dissertation.

Atlas Occulta was founded with the mission of promoting books in the Paranormal, Hermetica & the Occult Science. Our vision is to preserve the legacy of literary history by reprint editions of books which have already been exhausted or are difficult to obtain. Our goal is to help readers, educators and researchers by bringing back original publications that are difficult to find at reasonable price, while preserving the legacy of universal wisdom..

A book that should not be missing from the library of any serious student of the ancient alien question.

The University of Chicago

HORUS IN THE PYRAMID TEXTS

A DISSERTATION

SUBMITTED TO THE FACULTY OF THE GRADUATE SCHOOL OF ARTS
AND LITERATURE IN CANDIDACY FOR THE DEGREE
OF DOCTOR OF PHILOSOPHY

(DEPARTMENT OF SEMITICS)

BY
THOMAS GEORGE ALLEN

A Private Edition
Distributed By
The University of Chicago Libraries
1916

PREFACE

During my studies at the University of Chicago I have become deeply indebted to various members of the Semitic faculty: Messrs. Price, Willett, J. M. P. Smith, Luckenbill. To Professor Breasted in particular, under whom all the work in my principal subject has been done, I owe thanks not only for the stimulus imparted in his classes and for the use of his Pyramid Texts manuscript, but also for the inspiration of close personal contact and friendship.

THOMAS GEORGE ALLEN

CHICAGO
1915

TABLE OF CONTENTS

	PAGE
INTRODUCTION	9
SYMBOLS AND ABBREVIATIONS	14
HORUS IN THE PYRAMID TEXTS	15
Occurrences of the Name. *See* Appendix	73
Forms	15
Classified References	16
A. Epithets of Horus	16
B. Magical or Mystic Names of Horus	21
C. Relationships of Horus	21
I. Genealogical	21
II. Position Occupied by Horus in Relation to Other Divinities	22
III. Relations to Horus on Part of Other Divinities	23
a) Actions of Other Divinities toward Horus	23
b) Position Occupied by Other Divinities in Relation to Horus	24
c) Attitude of Other Divinities toward Horus	26
D. Nature of Horus	27
I. Attributes and Powers	27
II. Habitat	28
III. Attitude of Horus toward Other Divinities	32
IV. Parts of Body	33
V. Elements or Phases of Personality	34
VI. Subordinates	34
VII. Equipment	36
E. Activities of Horus	39
I. Involving Divinities Other than King	39
II. Involving King	40
a) Favorable	40
b) Unfavorable	47

PAGE

F. Eye of Horus 47

I. Epithets of the Eye 47

II. Magical or Mystic Names of the Eye 48

III. Relations of Horus to the Eye 48

IV. Relations of Others to the Eye 51

V. Actions, Circumstances, and Qualities of the Eye . . 56

VI. Parts and Accessories of the Eye 58

VII. Symbolism of the Eye 59

VIII. Two Eyes of Horus 62

IX. Unrestored Fragments Alluding to the Eye 63

G. Other Mythological References to Horus 63

H. Miscellaneous References to Horus 66

SUPPLEMENT: OFFSPRING OF HORUS 67

APPENDIX: OCCURRENCES OF DIVINE NAMES IN THE PYRAMID TEXTS 70

INTRODUCTION

In a group of five pyramids at Sakkara, the tombs of Fifth- and Sixth-Dynasty pharaohs of Egypt who ruled from about 2650 to 2475 B.C., is preserved the largest body of inscriptions known anywhere from that remote age. These mortuary and religious texts not only present through their allusions an invaluable commentary on early Egyptian civilization, but they reveal the earliest discernible phases of man's thoughts on the great problems of cosmic origin and human destiny. Their separate elements prove upon examination to have been composed under varying circumstances by which some portions can be dated as early as the predynastic age before the Delta was conquered by the Southland under Menes and the Two Lands thus finally united into one nation. But varying conceptions had already been so thoroughly amalgamated that the vicissitudes through which originally local beliefs and cults had passed are, it would seem, untraceable. It was to facilitate such progress as may, however, be possible along this line, that I at first planned to segregate and classify all references to all the deities mentioned in the Pyramid Texts.

The immediate requisite was a translation of the documents. Maspero's pioneer text, and with it many of his interpretations,[1] had been superseded by the new text edition of Sethe,[2] which appeared in 1908–10. The latter had autographed a preliminary translation with his text as previously transcribed for the monumental Berlin Dictionary of the Egyptian language, which has been in course of preparation by the four great academies of Germany since 1897. Another independent translation, based on the new text, but also preliminary, and in manuscript only, except for quotations used in his volume, had been made by Professor Breasted in preparation for the Morse Lectures which he delivered in 1912.[3] Using this latest translation as a guide, and with constant comparison of kindred

[1] Published first serially in *Rec. de Trav.*, then together as *Les Inscriptions des pyramides de Saqqarah.* Paris, 1894.

[2] *Die altägyptischen Pyramidentexte, nach den Papierabdrücken und Photographien des Berliner Museums neu herausgegeben und erläutert von Kurt Sethe.* Leipzig, 1908–10. The *Erläuterungen* have not yet appeared.

[3] Published under title of *Development of Religion and Thought in Ancient Egypt.* New York, 1912.

elements, I then made for myself a complete version of the Pyramid Texts.

All occurrences of divine names and epithets were next listed. Differentiation of independent deities from the mass proved a rather arbitrary matter. Not only do various types of supernatural beings appear, from the great cosmic powers down to representatives of the animal and vegetable world,[4] geographic incarnations,[5] and even inanimate objects,[6] but epithets or apparent epithets often stand in place of names. It was finally decided to consider primarily:

1. Terms regularly used as principal designation of well-known deities.

2. Terms apparently used as principal designation of less-known deities, whether traceable to epithetical origin[7] or not.

3. Names of celestial bodies, animals, and serpents, appearing as supernatural agents.

The following classes were among those set aside:

1. Group-designations.

2. Names of barques and crowns (unless supernatural serpent-element be involved).

3. Terms whose formation or use suggests merely epithetical function, whether or not the deity to whom they apply be determinable. Two special types of this class are: (*a*) epithets in form of prepositional phrases; (*b*) epithets used as mystic names only.

An index of all occurrences in the Pyramid Texts of divine names selected on the foregoing basis forms an appendix to this dissertation.

With the excessive wealth of material which even this selected list revealed, the original plan was, however, found to be too comprehensive for the purposes of a thesis. The god Horus, then, or rather the group of Horuses, has formed the main subject of investigation for the present.[7a]

In arriving at the translations which lie back of the Horus-citations, it became painfully evident that, apart from their archaic character, the Pyramid Texts, though engraved for the king himself, are by no means free from the textual corruptions so common in later

[4] E.g., serpents, *knmw·t*-bird, *y͗m*- and *nbś*-trees of Pyr. 808, etc.

[5] E.g., *Ymn·t*, "the West," in Pyr. 282 and 284.

[6] E.g., the Northern crown under various names in Pyr. 196.

[7] As e.g., *Ḏḥwty*, *Nḥb-k͗·w*, *Ḫnty-ymnty·w*.

[7a] G. Van der Leeuw's more general study of the *Godsvoorstellingen in de oudaegyptische pyramideteksten* (Leiden, Brill, 1916) came to my attention while my own work was in the press.

days. The signs *k* and *nb* are often confused,[8] but fortunately cause little difficulty. Dittography[9] or omission[10] of even whole phrases appears; sometimes even one element of a pun is lost.[11] Again, just as the pointing has occasionally obscured the sense of a Hebrew passage, wrong determinatives have sometimes found a place in these pyramid copies of more ancient texts.[12] More serious corruptions occur, such as that of *Tꜣy·t* in Pyr. 2074 (cf. 1794), where the parallelism of two or more bits of text is not so readily discoverable or where no parallel exists. To such scribal errors as these are probably due many of the passages which now defy translation.

Accurate interpretation is made difficult also by the fact that personal pronouns are often used ambiguously; and, worst of all, the being to whom a pronoun refers may even be left entirely unnamed,[13] so that the reader naturally connects the pronoun with the last suitable noun preceding, until some parallel phrase is noticed which reveals his error.

Through the grouping together of like statements in this dissertation and in contemplated future treatment of other gods besides Horus, textual criticism may, it is hoped, be somewhat assisted; but the primary effort has been to set down the tenor of the texts as they now stand.[14]

The solar element in Horus clearly predominates. The only Horuses named in as many as five Pyr. paragraphs are:

	Number of Paragraphs
1. Harakhte	30
2. *Ḥr šsmty*	11 ?
3. Eastern Horus	10
4. Horus of Dewat	9
5. Horus of the Gods	7
6. *Ḥr špd*	5 ?

[8] E.g., Pyr. 2075.

[9] E.g., Pyr. 385P.

[10] Cf. Pyr. 829 with 836.

[11] E.g., Pyr. 51cW.

[12] Certainly so, for example, in one of the texts of Pyr. 1227 (F 97) and in 21b; probably so in 472 (E 112).

[13] Certainly so in Pyr. 249 (F 153), cf. 614 (F 154); probably so in parts of the offering ritual, cf. F 106 ff.

[14] A detailed outline of the classification is included in the Table of Contents. The grouping of citations in the various subdivisions depends chiefly on alphabetic order of Egyptian key-words, in connection with sequence of paragraphs in Sethe's text. As the headings cannot be made mutually exclusive, cross-references to more extended statements are frequent. The king, whether representing Osiris or not, is treated independently; and, where the key-words are divine names, the king comes last.

Breasted[15] has already indicated the celestial nature of Nos. 1, 2, 3, and 5. Dewat too in this age is a quasi-celestial region,[16] and *Ḥr špd* in the Pyramid Texts is mentioned in celestial connections exclusively. The less common epithets of Horus are likewise largely celestial, as is his habitat.

On the other hand, the genealogy of Horus is almost as wholly Osirian as his physical nature is human. It is, however, stated that Horus, son of Osiris, is also son of Hathor;[17] *Ḥr špd* came forth from the king (=Osiris) and Isis-Sothis;[18] and the cloud-lord Orion, once called father of Horus,[19] is in Pyr. 819 identified with Osiris and is several times[20] connected with him. These few cases form but a slender link between the celestial Horus and the terrestrial Osiris; it must be noted too that the Orion connection is obtained only by moving Osiris skyward.

The most detailed myth traceable is that which records how Horus was born to Isis in Khemmis, fought with Set in his young manhood, and after recovering his eye,[21] which Set had taken and swallowed, bestowed it upon his father Osiris. The judicial proceedings which ensued at Heliopolis seem to have been a trial of Horus himself,[22] or again to have concerned the eye.[23] They resulted in any case in the defeat of Set.

Hostility between Horus and the dead appears in only three incidents: The king comes to Nut, having shaken off Horus behind him;[24] Re does not let Horus and Osiris gain control over the king's heart;[25] and one Horus warns away from the king's pyramid the protégés of another Horus, a comrade of Set, Osiris, etc.[26] Elsewhere, Horus is rather called upon to aid the dead king, not only by such offices as the myth assigned him on behalf of Osiris, but in many purely solar connections, especially noticeable in the sections on purification,[27] ascent,[28] deification,[29] and rule.[30] Moreover, the celestial Horus

[15] *Development of Religion and Thought*, pp. 154 ff.

[16] Breasted, *op. cit.*, p. 144, n. 2.

[17] C 10. [18] C 19. [19] C 11. [20] Pyr. 186, 882–83, 925, 959–60.

[21] A solar touch in Pyr. 670 (F 42) is that Horus "circled about ⌜in search of⌝ his eye."

[22] G 44, 54.

[23] F 72, 48, 59, 100. On the judgment see also G 57–58.

[24] Pyr. 250 (C 72).

[25] Pyr. 145–46 (C 59; E 178–79).

[26] Pyr. 1264–74 (E 180–81).

[27] E 94 ff. [28] E 108 ff. [29] E 121 ff.

[30] E 142–45; note also E 150, 160, 166, 170, 173.

is utilized in magic rites: "charms of Harakhte" must be learned;[31] Re and Horus together are involved in a serpent-charm;[32] and Horus, Isis, and Atum restrain a (celestial?) bull (from harming the king).[33]

The actual god might not be present to assist the dead. But he was in many cases impersonated by the officiating priest. Parallel to the instances[34] in which Horus is said to bestow his eye (representing various offerings) upon the king are others in which an unnamed "I" (presumably the priest) likewise presents the eye of Horus.[35] The king is here regularly called Osiris. This "I" occasionally calls himself "Horus"[36] and even "thy son Horus."[37] So the Osirian relationship of father and son was re-enacted between the participants in the mortuary ritual.

As the priest's services for the dead symbolized those of Horus for his father Osiris, so the eye of Horus, the gift of which was involved in the myth, was first transformed into a symbol for any form of mortuary offering.[38] Then the mortuary functions of Horus were delegated to his eye in its various capacities: the eye could avenge the deceased, help him ascend, etc., as had Horus himself.[39] The eye of Horus was further identified with the sun,[40] and, like Horus himself, endowed with human form.[41] Lastly, the deceased king's own person, so often identified with Horus, became identified likewise with the magic Horus-eye.[42]

31 Pyr. 855–56 (D 223).

32 Pyr. 673 (D 249).

33 Pyr. 1302 (D 151).

34 Summarized in E 176.

35 F 214–21.

36 Pyr. 11 and 905.

37 Pyr. 69.

38 See summary after F 423.

39 See F 228 ff.

40 Pyr. 698N (F 280), either an instance of identification of Re and Horus or a further case of the eye assuming place originally belonging to Horus himself.

41 Pyr. 2088 and 2090 (F 318–19).

42 F 130.

SYMBOLS AND ABBREVIATIONS

Arabic numbers cited alone or following the word "Pyr." refer to the paragraphs in Sethe, *Die altägyptischen Pyramidentexte,* corresponding also to the *Zettel*-numbers of the Berlin Dictionary. Small letters occasionally following these numbers indicate individual lines of Sethe's paragraphs. The capital letters W, T, P, M, and N sometimes added are conventional abbreviations for the names of the builders of the respective pyramids; they serve to distinguish parallel texts. To avoid confusion, no citations have been made by *Spruch*-numbers.

Arabic numbers preceded by one of the capital letters A–H refer to other portions of the thesis itself.

Titles of books and periodicals are, it is thought, cited with sufficient fulness. Where Breasted is quoted without a title following, reference is to his manuscript translation of the Pyramid Texts.

Egyptian words (in italics) are transliterated as in modern German treatises, except that the two signs *ỉ* and *j* have both been replaced by *y*.

[ ] indicates restoration.
⸢ ⸣ indicates uncertainty.
( ) indicates explanatory additions by the author.
- indicates construct relation or compound preposition.
• separates formative elements of words.
—represents each word lost, up to five words.
——— represents more than five words lost.
. . . is sometimes inserted to mark omitted portions of a phrase.

HORUS IN THE PYRAMID TEXTS

OCCURRENCES OF THE NAME

See *Ḥrw* in Appendix.

FORMS

except in the following cases:

in W only, in 18, 58–60, 64, 72, 73, 76–81, 83–87, 91, 94, 95, 97–100, 115, 304, 487.

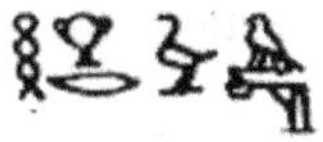

in M only, in 1686 and 1690; may not be Horus.

dual, in 457W and 695TN.[a]

nisbe fem. pl., in 487WN, 598PN, 770PMN, 961PMN, 1735N, 1928N, 2099N.

nisbe fem. pl., in 943N.

nisbe fem. pl., in 598T.

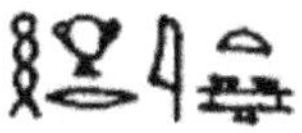

nisbe fem. pl., in 943PM.

[a] Cf.

(*Ḥr·wy-mš*) as n. pr. in *Aeg. Inschriften aus den k. Museen zu Berlin*, II, 318. *Ḥr·wy* occurs also on a Ptolemaic stela published by Touraieff in *Rec. de Trav.*, XXXVI, 72; this is clearly *not* a writing of Atum as assumed in the review in *Ancient Egypt*, 1915, p. 26. That *Ḥr·wy* represents two separate gods is indicated in the latter case by the deceased lady's epithet *s·ḥtp·t nṯr·wy*.

CLASSIFIED REFERENCES

* Not directly dependent on the name Horus.

A. Epithets—

1 *ꜣw-nmt·t*, "long-striding." *853(D 2)

2 *yꜣbty*, "eastern."[44] 450(C 73, D 28), 527 (D 46), 982(D 48), 1085(D 34), 1087(C 127), 1132(D 49), 1258(C 51), 1410(D 47), 1414(D 66), 1478 (D 11, D 124)

3 *yꜣḫty*, "of the horizon."[45] 4(D 42), 7(H 2), 337(D 31, D 36), 342(D 35), *346(C 35), 348(C 58), 351(D 32, D 36), *353(D 44), 358(D 32, D 36), 360(D 38), 526(D 46), 855–56(D 223), 891P(D 57), 926(D 32), 927(D 37), *928(D 44), 932(D 32), 933(D 37), *934(D 44), 1049(D 60), *1085(D 34), 1087(C 127), 1103(D 33), 1384(H 9), 1411(D 47), 1415(D 66), 1449 (E 148), 1478(D 11, D 124), 1693(C 57, B 9)

4 *ywꜥ-yt·f*, "heir of his father." 316(D 7, G 54)

5 *ymy Ysyr*[46] *X*, "(dweller) in Osiris X (king's name)." 19a(F 43), 21b(F 63), 55(F 43, F 77), 831(F 43)

6 *ym(y)·w(y) ꜥḥ*, "(dwellers) in the palace,"[47] epithet of Horus and Set. *141(D 113)

7 *ymy wꜣḏ wr*, "(dweller) in the great green (sea)." 1505(C 83)

8 *ymy wsḫ·t*, "(dweller) in the broad hall." 905(E 90)

9 *ymy byk·w*, "(dweller) among the falcons." *⌜1672⌝(E 164)

10 *ymy pr·f*, "(dweller) in his house."[48] 1294(D 194)

11 *ymy Špd·t*, "(dweller) in Sothis." 632 and 1636(B 6)

12 *ymy Ḏbꜥ-ḫrw·t*, "(dweller) in *Ḏbꜥ-ḫrw·t*." *734(D 108)

13 *ꜥnḫ*, "living."[49] 1807

14 *ꜥnḫ-ḫꜥ·w*, "living of dawnings," Horus-name of Mernere. 7, 8

15 *wꜣḏ·wy*, "green," epithet of *Ḥr·wy*, dual.[50] 457(H 4)

16 *wꜣḏꜣḏ*, "⌜bright⌝ green (bird)," an aspect of *Ḥr Dwꜣty*. 1207(D 61)

17 *wꜥ*, "sole one."[51] *853(D 2), *854

[44] Possibly this god is meant by *nb-yꜣb·t*, "lord of the East," in 1486.

[45] Cf. pl. in 1155. A god called *ꜥnḫ m yꜣḫ·t*, "he who lives in the horizon," is mentioned in 151.

[46] The old readings of Osiris and Isis with prosthetic *y* (Erman, *Glossar*, p. 16) are retained for convenience, though Erman has since shown (*Zeitschrift für äg. Sprache*, XLVI, 92) that name of former began with *w* and was probably either *Wsry* or *Ws-yry*, while for Isis he now (*Gram.*[4], Schrifttafel Q 1) gives *ꜣs·t*.

[47] Cf. 598(D 82). The reading *ꜥḥ* for "palace" appears most clearly in Pyr. 141 and 585. It was accepted by Erman in 1912 (*Die Hieroglyphen*, p. 32).

[48] Applied to Min in 1993(C 105).

[49] Cf. *ꜥnḫ* as son of Sothis in 458 and 1482P (MN have *ny-ꜥnḫ*); *ꜥnḫ m yꜣḫ·t*, "him who lives in the horizon," a god otherwise unnamed in 151; and *sbꜣ ꜥnḫ ḫnty sn·w·f*, "the living star presiding over his brothers," with whom king is compared in 904.

[50] Cf. *wꜣḏ pr m Wꜣḏ·t*, "green one who came forth from (the goddess) Buto," in vocative in 1875.

[51] Cf. *wꜥ*, epithet of king as *kꜣ n p·t*, "bull of the sky," in 293; *wꜥ pn ḏd rꜥ nb*, "this *one* who endures every day," used of some solar divinity in 1078; *nb wꜥ*, "the sole lord," perhaps applying to *kꜣ-Nḫn*, "the bull of Hieraconpolis," in 276; *ḥry wꜥ·f*, "him who is by himself," a solar divinity in 309.

A. Epithets—

18 *wr (y)m(y) Ywnw*, "great one (dwelling) in Heliopolis," epithet of *Ḥr·wy*, dual.[52] 695(H 5)
19 *wr pn*, "this great one."[53] *103(F 425), *583
20 *wr s3-wr*, "great one, son of a great one." *852(G 27)[54]
21 *b3 ymy dšr·f*, "soul (dwelling) in his blood." *854(D 55)
22 *byk nṯry*, "divine falcon," an aspect of *Ḥr Dw3ty*. 1207(D 61)
23 *pr m Ḥʿpy*, "he who came forth from the Nile." *2047(C 79)
24 *pr m šnṯ*, "he who came forth from the *šnṯ*-serpent." 681(C 80)
25 *pr m šnḏ*, "he who came forth from the acacia." *436(G 50)
26 *mryy-t3·wy*, "beloved of the Two Lands," Horus-name of Pepi I. 6,[55] 7
27 *mḥty*, "northern." 1295(D 88)
28 transliteration and translation uncertain; form below. 1863(C 104)

29 *ny-šw·t-nṯr·w*, "king of the gods." 1458(E 123)
30 *nb-y3ḫ·t*, "lord of the horizon," epithet of Harakhte.[56] *7b
31 *nb-w3ḏ*, "lord of the green (cosmetic?)." 457(D 24)
32 *nb-p·t*, "lord of the sky."[57] 888(D 5)
33 *nb-p·t*, "lord of the sky," epithet of Harakhte. 7b
34 *nb-pʿ·t*, "lord of men." 14(E 86), 737(F 149), 1258(C 51), 1804(E 146)
35 *nb-pʿ·t nṯr·w*, "lord of men and gods." 895(C 99)
36 *nb-m3k·t*, "lord of the ladder."[58] *974, *980(D 183)
37 *nb-t3·wy*, "lord of the Two Lands." 1258(C 51)
38 *nb-ḏf3·w*, "lord of food," epithet of *Ḥr·wy*, dual. 695(H 5)
39 *Nḫny*, "of Hieraconpolis."[59] 295–96(D 236), 1293(E 98, D 194), 2011 (E 141)
40 *nḫn ḫrd*, "young, child." 1320(C 49)
41 *nṯr*, "god." *971, *974–75(D 252),[60] *978–80(D 253), *1863(C 104)
42 *nṯr·w*, "of the gods." 525(D 46), 981(D 48), 986(D 65), 999(D 39), 1086(D 34), 1408(D 47), 1412(D 66)
43 *nṯr ʿ3*, "great god." *[70](F 427)

[52] Cf. *ymy Ywnw*, "(dweller) in Heliopolis," a magical name applied to king in 181. In 716 king is identified with *k3-Ywnw*, "the bull of Heliopolis."

[53] *Wr pn* is applied to king as Osiris in 777–78, 1796.

[54] Cf. 853b and 2032. *Wr* = Geb in 1702.

[55] Restore

in the break; cf. Pyr. 7a and *Urkunden des äg. Altertums*, I, 94:16.

[56] Occurs independently in 277, 409, 1172.

[57] Cf. *nb-p·t* apparently equated with Osiris in 964–68.

[58] Dependent on *nṯr;* for connection with Horus cf. 973b.

[59] Cf. *k3-Nḫn*, "the bull of Hieraconpolis," in 276.

[60] Cf. 973b with 974a.

A. Epithets—

44 *nṯr ꜥꜣ*, "great god," epithet of Harakhte.[61] *7b

45 *nṯr-ḫꜥ·w*, "god of dawnings," Horus-name of Pepi II. 7, 8

46 *nṯr śmśw*, "first-born god," epithet of Horus son of Osiris. *466(C 89)

47 *nṯr·wy*, "the two gods," epithet of Horus and Set.[62] *1148(C 69)

48 *nṯr·wy ypw wr·w(y) ꜥꜣ·w(y)*, "these two great and powerful gods," epithet of Re and Horus.[63] *952(C 110)

49 *nṯr·wy ꜥꜣ·wy*, "the two great gods," epithet of Horus and Thoth.[64] *1571(E 144)

50 *nḏ yt·f*, "he who avenged his father." 633 and 1637(B 8), 1685(E 74)

51 *nḏ yt·f Yśyr*,[46] "he who avenged his father Osiris." 898(C 102), 1406N (D 204)

52 *rw*, "lion."[65] *436(G 50)

53 ⸢*rpꜥty*⸣, "hereditary prince." 1458(E 123)

54 *rnpy*, "youthful." 25c and 767(E 17)

55 *rḫ špś·w-nṯr*, "intimate of the worthies of the (Sun-)god." *815PM

56 *rśw*, "southern." 1295(D 88)

57 *ḥwnty*, "⸢he of the maiden (or 'pupil [of the eye]'?)⸣."[66] 206(D 274)

58 *ḥry-yb Yrw*, "(dweller) within *Yrw*."[67] 723(D 93)

59 *ḥry-yb Dwꜣ·t*, "(dweller) within Dewat." 1959(F 209)

60 *ḥry šdšd-p·t*, "upon the *šdšd* of the sky."[68] 800(D 51), 1036(D 52)

61 *ḥry-ḏꜣḏꜣ m·ꜥnḫ·t·f n·t mꜣꜥ·t*, "master of his sustenance of truth."[69] 815(E 114)

62 *ḥry-ḏꜣḏꜣ rḫ(y)·t·f*, "master of his people." 644(C 81)

[61] Is applied also to Osiris in 465(G13), Geb in 1616, Re in 1471, and an otherwise unnamed solar deity in 1208(G 32).

[62] Dual *nṯr·wy* without epithet occurs also in 273, 903.

[63] Same phrase occurs once more, in 1690(D 264, cf. C 57).

[64] Same phrase occurs also in 467, 1125–26, 1253, 1738, 1750, [1985]. Cf. also *nṯr·wy ypw ꜥꜣ·wy* in 1010.

[65] Cf. *rw* in 422 and 426, unintelligible; *pf rw* and *pn rw*, referring possibly to Horus and Set, in 425; and phrase *rw ḥꜣ rw n ꜥnḫ*, "(one) lion (is) behind (another) lion for life," in 690. Horus is found wearing the lion-mask in 973(D 234).

[66] The noun *ḥwn·t* itself occurs literally as "maiden" in 809 and ⸢1487⸣ and in *ḥwn·t wr·t ḥry·t-yb Ywnw*, "the great maiden (dwelling) within Heliopolis," in 728 and 2002; and figuratively as "pupil (of the eye)" in 93a(F 302) and ⸢682⸣. Cf. also *ḥwr·t wr·t . . . ḥry·t-yb Nḫb*, "the great *ḥwr·t* . . . (dwelling) within El Kab," in 2204a. Möller (*Über die in einem späthieratischen Papyrus des Berl. Mus. erhaltenen Pyramidentexte*, p. 23) suggests that this may be error for *ḥwn·t*; but *ḥwr·t wr·t* is represented by *śmꜣ·t wr·t*, "the great wild cow," in the parallel text in 729a. May we understand that Horus is son of the maiden of Heliopolis?

[67] *Yrw* occurs again in 864: king's waters came from Elephantine, his *nṯry*-natron from *Yrw*, his *ḥsmn*-natron from the Oxyrynchite nome, and his incense from Nubia (*tꜣ sṯ*); possibly also in 456a: "Sobk, ⸢(my)⸣ lord, ⸢soul of *Yrw*⸣" (or 'Sobk, lord of *Bꜣyrw*'?).

[68] In 539a and 540a king ascends to sky upon the *šdšd* which is in the zenith (*ymy wp(w)·t;* latter written in full in 854c, where Breasted so translates it).

[69] In 1483 the offspring of Horus are said to live on truth (*ꜥnḫ m mꜣꜥ·t*).

A. Epithets—

63 *Ḫm(y)*,[70] "of Letopolis." 2078(C 26)

64 *ḫnty yꜣḫ·w*, "presider over the spirits." 800(C 82), 1505(C 83), 1508(C 84), 1518(D 12)

65 *ḫnty ꜥnḫ·w*, "presider over the living." 2103(C 103)

66 *ḫnty pr·w*, "presider over estates."[71] 133c(C 61, D 4)

67 *ḫnty·wy pr·w*, "presiders over estates," epithet of *Ḥr·wy*, dual. 695(H 5)

68 *ḫnty mn·wt·f*, "presider over his ⸢thighs⸣" (cf. det. in 804).[72] 804PMN and 1015P(C 85)

69 *ḫnty mny·t·f*, "presider over his ⸢—⸣."[73] 1015N and 1719M[N](C 86)

70 *ḫnty Ḫm*, "presider over Letopolis."[74] 810(D 1, G 55)

71 *ḫnty y·ḫm·w sk*, "presider over the imperishable ones," epithet of Horus of Dewat. 1301(D 3)

72 *ḫnty ḫnty sḫm·w*, "presider presiding over the mighty." *1294(D 194)

73 *ḫnty D(w)ꜣ·t*, "presider over Dewat."[75] 5b(D 72)

74 *ḫsbḏ yr·ty*, "blue-eyed." 253(E 11)

75 *Ḫꜣty*, "of *Ḫꜣ·t* (a city)." 1257N(C 87)[76]

76 *ẖrd nḫn*, "the young child." 1214(G 3)

77 *ẖrd nḫn ḏbꜥ·f m rꜣ·f*, "the young child with his finger in his mouth." 663(D 292), 664(C 88)

78 *sꜣ nḏ yt·f*, "the son who avenged his father." 633 and 1637(B 8)

79 *šꜣꜣ*, "sated." *854

80 *sbꜣ wpš p·t*, "the star that illumines the sky," applied to Horus (of Dewat).[77] *362(D 45)

[70] Value *ḫm* is now assigned to

by Erman (*Gram.*[4], Schrifttafel O 74). It is indicated by such spellings as

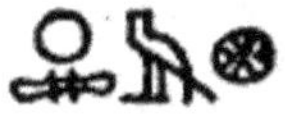

(Pyr. 1670aN) and

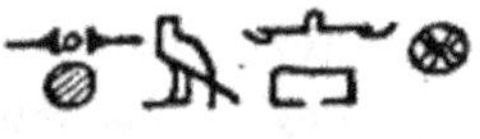

(Piankhi 11:6 = *Urkunden des äg. Altertums*, III, 46). The Piankhi passage makes Horus the "lord (*nb*)" of Letopolis.

[71] Occurs in throne-name of Amenhotep III on a granite column in British Museum, No. 64 (Budge, *Book of Kings*, I, 139). Cf. the *yꜣḫ·w-pr·w*, "house-spirits," in 842 whose purity comes in connection with Horus's purification of the king in 841(D 54). In 1881 *nb-pr*, "lord of the house," is an epithet of the king.

[72] Same epithet represents a god otherwise unnamed in 285T and 655TMN. In 1549P it is given to Osiris, while in 1552P it seems to belong with *Šsmw*.

[73] Same epithet represents an otherwise unnamed divinity in 285W.

[74] Same epithet appears independently in 419, 908, 1175, 1723.

[75] Same epithet represents an unnamed god with whom king is identified in 715. Cf. also a *nb-⸢mr⸣·w D(w)ꜣty·w*, "lord of the Dewat-lakes," in 1530.

[76] P has the older ambiguous writing *Šꜣty*.

[77] Same epithet represents a divinity with whom king is identified in the similar text 1455. In 1038 king ascends to sky as *sbꜣ ꜥꜣ ḥry-yb yꜣb·t*, "the great star (dwelling) within the East"; in 904 king is compared with *sbꜣ ꜥnḫ ḫnty sn·w·f*, "the living star presiding over his brothers"; in 1048 king is adorned as *sbꜣ wꜥty ḥry-yb Nw·t*, "the single star (dwelling) within Nut."

A. Epithets—

81 *šbꜣ dꜣ wꜣḏ wr*, "the star that ferried over the great green (sea)," epithet of *Ḥr ḫnty yꜣḫ·w*. 1508(C 84)

82 *špd*, "the ready." 330W and 331T (latter *špd·t!*) (C 112, D 73), 632 and 1636(C 40), *⌜1863⌝(C 104)[78]

83 *šmšw*, "the first-born."[79] 301(D 263)

84 *šḫm m yꜣḫ·t*, "mighty over the horizon." *853(D 2)

85 ⌜*šḫm m*⌝ *Nbty*, "⌜mighty over⌝ him of Ombos,"[80] used after name of Horus as a royal title, the whole written as below. 6–8, 786

86 *šḫm m nṯr·w*, "mighty over the gods." *853(D 2)

87 *Škšn*, "⌜—⌝."[81] 1734(C 93)

88 *Šꜣt(y)*, "of *Šꜣ·t* (a city)." 450W(C 73, D 28), 1257P(C 87)[82]

89 *šn m sꜣ-yr·t·f*, "he who went about in the protection of his eye." 195(C 92), 198(C 1, C 92)

90 *šsmty*, "⌜—⌝."[83] *342(D 35), 450(C 73, D 28), *⌜456⌝(D 155), 528(D 46), 983(D 48), 987(D 65), 1085(D 34), 1136(D 49), 1409(D 47), 1413(D 66), 2062(D 63)

91 *tp(y) yḫm·wt*, "upon (or 'at the head of' ?) the *yḫm·wt*." 1951

92 *tp(y) nḏm·w šṯ(y)*, "at the head of the sweet-smelling ones," epithet of *Ḥr ḫnty yꜣḫ·w*. 1518

93 *y·tm whm kꜣ·t·f*, "not repeating his work." 1622(C 76)

94 *ṯmšty*, "⌜the gleaming one⌝."[84] 702(C 113)

95 *Dwꜣ-nṯr*, "Morning-star," an aspect of Horus of Dewat. *1207(D 61)

96 *D(w)ꜣty*, "of Dewat."[85] 148(D 180), *362(D 45), *802(E 108), 877(C 114), 1134(D 49), 1207(D 61), 1258(C 51), 1301(D 3), 1734(C 93)

97 *dšr yr·ty*, "red-eyed." 253(E 11)

98 *dšr·t tpy·t Rꜥ*, "the red crown that is on the head of Re," epithet of *Ḥr ṯmšty*. 702(C 113)

99 *Ḏbꜥ-ḫrw·t*, "of *Ḏbꜥ-ḫrw·t* (a city)." 1993(C 105)

[78] *Ḥr špd* as a unit is everywhere in Pyr. written without the determinative

regularly found (as in 1863) with the independent divine name *Špd*.

[79] Cf. 309 and 313: the king, in 308 and 312 equated with Osiris, is in 309 identified with *ḥry wꜥ·f*, "him who is by himself." The parallel text in 313 has instead *ḥry rd·f*, "him who is upon his foot." In both cases the god is called *šmšw-nṯr·w*, "first-born of the gods," and is a solar divinity. Again in 306 the king, who seems to have become a solar Horus (cf. 304), is called *šmšw r wr*, "more first-born than the eldest."

[80] I.e., Set; cf. 204a.

[81] *Škšn* occurs as an independent divine name in ⌜498⌝, 1440, ⌜2186⌝.

[82] N has the more exact *Ḫꜣty*.

[83] Fem. *šsmty·t* is epithet of Sekhmet; cf. 262b and [2206].

[84] This root occurs also in 911, 1147, 1349, 1460. In 911 *ṯmš·t dšr·t* with Northern crown on basket as det. is "the bright red crown" (so Breasted). In 1147 king is *ššd pw n ṯmš·wt*, "this band of ⌜bright⌝ (colors)." In 1349 the divinity *Bꜣby* is called *dšr-msḏr ṯmš-ꜥr·t*, "with red ear and ⌜glistening jaw⌝." (On *ꜥr·t* as "jaw" cf. determinatives in Pyr. 30 and Budge, *Book of Opening the Mouth*, II, 162.) In 1460 king is *nw n ṯmšš·t*, "this ⌜brightness⌝" which came forth from ⌜Nun⌝.

[85] Cf. *ymy D(w)ꜣ·t*, "dweller in Dewat," in 330T and 331W(D 73).

B. Magical or Mystic Names—

1 *yꜣḫ ymy ḏndrw*, "spirit (dwelling) in the *ḏndrw*-barque," used of *Ḥr špd*. 633, 1637
2 *ym(y) ḥnw*, "(dweller) in the *ḥnw*-barque (Sokar's)." 138
3 *nwr·w*[86] *n·f tꜣ šdꜣ·w n·f p·t*, "he before whom earth quakes and the sky trembles." 143
4 *Rꜥ*, "Re," used of *Ḥr Šꜣt(y)*, *Ḥr šsmt(y)*, and Eastern Horus treated as one god. 452
5 *ḥnw*, "*ḥnw*-barque (Sokar's)." 620
6 *Ḥr ymy Špd·t*, "Horus (dwelling) in Sothis," used of *Ḥr špd*. 632, 1636
7 *Ḥr bꜣy·t-rp·t*, "Horus, ⸢—⸣ of the *rp·t*-chapel."[87] 767
8 *Ḥr sꜣ nḏ yt·f*, "Horus, the son who avenged his father," used of *Ḥr špd*. 633, 1637
9 *ḥr yr nṯr·w*, "distant from the gods," used of Harakhte. 1693
10 *ḥkꜣ-⸢šnṯr⸣*, "ruler ⸢of incense⸣,"[88] used of *Ḥr Šꜣt(y)*, *Ḥr šsmt(y)*, and Eastern Horus treated as one god. 452
11 *šp šꜣꜣw*, "blind ⸢in attack⸣,"[89] used of the Horus hostile to the dead. 1268
12 *Km wr* (or *ḥt·t-Km wr?*), "great black (⸢—⸣)."[90] 1658(D 18)

C. Relationships—

I. GENEALOGICAL

1 *Yḫ·t wr·t* (*Yḫ·t-wt·t* resp.) bore king as Horus who went about in protection of his eye (*šn m sꜣ-yr·t·f*). 198
2 Horus is son of Isis. 1214(G 3), 1640(G 4)
3 Horus is son of Osiris. 22(G 10), 146(E 179), 179(G 8), 465(G 13), 466(C 89), 795(G 39), 898(C 102), 969(E 122, E 140), 1129(E 5), 1331 (C 90)
4 Horus was born to Osiris, Set was conceived for Geb. 144

[86] Or *wr·w?* The *n*-wave might be genitive; and *wr* alone probably does mean "quake" in 516a. Besides, *nwr*, which is found also in 581, 789, 956, 1183, 1270, 1357, 1855, and 2109, shows an initial *nw*-jar or -adze everywhere else except in 1855aN. But in our same text (W) *rn·f ym(y) ḥnw*, without genitive *n*, occurs in 138c, only four lines back.

[87] *Bꜣy·t-rp·t* is a mystic name of king as Osiris in 580; *rp·t* alone (with fuller writing *rpw·t* in 1349 and 207) is a chapel (shown by word-sign in 580 and 767, used as det. in 823 and 1349), located in Heliopolis (823) or in Dendera (207).

[88] Budge's translation of

(*Liturgy of Funerary Offerings*, p. 46) as "Sovereign of the divine *Sa* (*Sa* being 'fluid of life')," for which he credits Maspero, results from a misreading of the third sign as

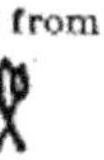

and failure to distinguish between *š* and *s*, as was still done in the Pyramid Age.

[89] Used as mystic name of the hostile *ḫꜣty·w* and *ymy·w yꜣw* in 1274(E 181). Our *šꜣꜣw* (spelled *šꜣw* in 1274) is found as verb *šꜣw* also in 651PM, where T shows the more exact radicals *ḫꜣw* used again in 588 and listed in Erman (*Glossar*, 99) with the meanings "etwas in etwas anreiben, einmischen."

[90] *Km wr* is a mystic name of king in 628 and 1630, is a god in 1390, and is an enemy of the king in 1350. The name is used elsewhere (outside of Pyr.) for the Bitter Lakes.

C. Relationships—

5 Osiris is father of Horus. 13(E 86), 146(E 179), 176(G 11), 472(E 112), 493(E 4), 573(C 101), 740(G 16), 758(G 14), 898(C 102), 1215(G 3), 1330(G 9), 1406(D 204), 1658, 1682(G 23), 1730(G 19), 1799(D 261), 1860(G 20), 1980(G 25), 2115(E 163)

6 Horus is son of Atum. 874(H 6), 881(D 56)

7 Horus is a great one, son of a great one (*wr sꜣ-wr*). 852[54]

8 Horus of Dewat and the king are children (*mś·w*) of Nut.[91] 1213(D 181)

9 Re calls himself brother ⸢of (Horus-)Soped⸣.[78] 1863(C 104)

10 Horus the son of Osiris is also son of Hathor. 466(C 89)

11 The cloud-lord (*nb ḳry*) Orion is father of Horus. 261(E 21)

12 Set is brother of Horus. 1742(F 226)

13 Horus is son of Geb. 973(C 91)

14 Horus the son of Osiris is seed (*mtw·t*) of Geb. 466(C 89)

15 Geb is father of Horus. 973(D 234), 977(F 59)

16 Horus is son of the king. 11(C 122), [44](C 64), 69(F 427), 192(E 40), 257(E 20), 578(E 72, F 50), 644(C 122), 1007(E 75), 1010(E 115), 1683(E 27), 1702(E 165), 1813(E 25), [1976](E 76), 1979(E 173, E 60), 1994(E 143)

17 King as Osiris is father of Horus. 101(E 35), 103, 258(D 138), 580(E 18), 589(E 18), 640(C 63), 650, 767(E 18), 1334(D 15), 1335(E 59), 1550(G 21)

18 King begets Horus. 143

19 *Ḥr špd* came forth from king and Isis-Sothis. 632, 1636

20 King begot (*mś*) (or 'is child of'?) Horus (dwelling) in *Ḏbꜥ-ḫrw·t*. 734 (D 108)

21 Horus was born to, and Set was conceived for, the king. 142

22 King was born to Horus, was conceived for Set, and received his purification in the Heliopolis-nome from his father Atum. 211

23 Horus had children (*mś*, *mś·w*).[92] 24(E 137), 619(E 101), 637(E 102, *E 10*),[93] 643(*E 2*), 766(E 137), 1338(*E 10*), 1548(*E 5*, D 136), [1823] (*E 10*, C 97), 1828(E 138)

24 King has joined himself with (*smꜣ·n·k ḥnꜥ*) the (issue) of body of Horus (*n·w ẖ·t·f*). 647

25 Horus had offspring. 647 (*mśyy·t* T, *mśw·t* PN, *mś·t* M) (G 31), 1333 (*mśyy·t* P) (*E 3*), 1339 (*mś·t* P) (*E 11*)

26 Horus of Letopolis had offspring. 2078 (*mśw·t* N) (*A 3*, *E 14*)

27 Horus the son of Osiris had an eldest daughter (dwelling) in *Ḳdm* (*sꜣ·t·f wr·t ymy·t Ḳdm*).[94] 1008(E 75)

II. POSITION OCCUPIED BY HORUS IN RELATION TO OTHER DIVINITIES

28 Belongs to Isis. 741(G 2)

29 Heir of his father. 316(D 7, G 54)

30 ⸢Mighty over⸣ him of Ombos. 6–8 and 786(A 85)

[91] Children of Nut are mentioned again in 823, where probably Horus and Set are meant.

[92] For complete references to the four Horus-sons, see Supplement.

[93] Italics here and following refer to the Supplement.

[94] But in 1977d (E 76) same phrase has pronoun *k* referring to king as Osiris. A ⸢related⸣ *Ḳdm·w* (pl.) occurs in a broken and unintelligible passage, 1970b.

C. Relationships—

31 For relation to *Nḥb-kꜣ·w*, see 346(C 35)

32 *Ḥsmnw*, Eastern Horus, Eastern Soul, and Harakhte, treated as one god, is one who oversees (*mꜣ ḥr-ḏꜣḏꜣ*) the gods (*nṯr·w*) without any god overseeing him. 1479

33 For relations to the gods, see also 301(D 263), 454(F 89), 466(C 89), 853(D 2), 895(C 99), 1458(E 123), 1693(B 9)

34 Re-Harakhte a compound divinity.[95] 1049(D 60)

35 King, like (*yś*) Har(akhte), is summoned by Re and receives offering from *Nḥb-kꜣ·w*. 346

36 For relations to Re, see also 452(B 4), 702(C 113), 2046(D 40)

37 For relation to ⸢the maiden⸣ (*ḥwn·t*), see 206(D 23)

38 For relations to Sokar, see 138(B 2), 620(B 5), 1823(C 97)

39 *Ḥr ḫnty mn·wt·f* (*mny·t·f* resp.) seems to be identified with *Śꜣṯwty nb-Śbw·t*. 804 and 1015(C 85)

40 *Ḥr špd* is in Sothis. 632 and 1636(B 6)

41 Horus-*Škšn* a compound divinity.[81] 1734(C 93).

42 Knows the worthies of the (Sun-)god (*špś·w-nṯr*). 815PM(A 55)

43 Heir of Geb. 1489(D 25)

44 For relation to Morning-star (*Dwꜣ-nṯr*), see 1207(D 61)

45 Is in Osiris the king. 19(F 43), 21(F 63), 55(F 43, F 77), 583, 831(F 43)

46 Is in embrace of Osiris the king . . . and escapes not from (*byꜣw r*) him. 636

47 Has not escaped from king. 1633

48 Is near (*ḫśf m*) king. 1234

49 Horus, youthful, child (*nḫn ḫrd*), is told of ascent of king to sky. 1320

III. Relations to Horus on Part of Other Divinities

a) *Actions of Other Divinities toward Horus*

50 Actions of *ꜣkr*. See 555(D 169)

51 Isis and Nephthys prevent that Eastern Horus, Horus lord of men, Horus of Dewat, and Horus lord of the Two Lands decay. 1258

52 After avenging king, Isis and Nephthys have given him to Horus. 584

53 Actions of Isis. See also 1140(C 96)

54 Osiris speaks to Horus when latter has plucked out evil [from king]. 1978

55 Actions of children of Horus ([*mś·w*]-*Ḥr*). See 1823(C 97)

56 Nut has taken Horus and Set and their sorceresses (*wr·t-ḥkꜣ·w*, = crowns[96]) to be her two eyes. 823

57 These two great and powerful gods (*nṯr·wy ypw wr·wy ꜥꜣ·wy*) who preside over the Field of Rushes (1690) cause that king live (again) and *nḥr* the seasons (*y·tr·w*) of Harakhte. 1693

58 "These four gods who stand (leaning) upon the *ḏꜥm*-staves of the sky" announce king to Re and Harakhte.[97] 348

[95] Cf. 1087–88(E 169, D 266).

[96] Cf. later examples, as at Medinet Habu, transl. in Breasted, *Ancient Records*, IV, § 66.

[97] Parallel text of 339 defines these gods more closely as "these four first-born spirits presiding over the curly-haired ones (*yꜣḫ·w šmś·w ḫnty·w ḥnskty·w*), who stand in the east side of the sky leaning (*dšr*) upon their *ḏꜥm* staves"; in 340 they announce the king to Re and *Nḥb-kꜣ·w*.

C. Relationships—

59 Re-Atum gives not king to Osiris nor to Horus. 145
60 Actions of Re. See also 351 etc. (D 32), 362(D 45), 1103(D 33)
61 *Ḫnty ymnty·w* brings food and offerings for *Ḥr ḫnty pr·w.* 133
62 Geb brings Horus to Osiris the king to aid latter. 590, 612, 634
63 Geb has caused that Horus see his father Osiris the king. 640
64 [Osiris the king] is bidden to put [his son] Horus within him. 44
65 King spreads out his arms behind Horus (i.e., embraces him as his ka, cf. D 192). 585, 636
66 King is bidden to loose Horus from his (cumbersome) regalia (*šṯ*[98]), that latter may punish followers of Set. 1285
67 King as Osiris is bidden to awake for (*rś n*) Horus and arise against (*ʿḥʿ r*) Set. 793, 1259
68 King is bidden to awake for (*rś n*) Horus and arise for (*ʿḥʿ n*) Set. 1710
69 King satisfies the two gods Horus and Set with a food-offering, so that they are satisfied and *špt.* 1148
70 Osiris the king is bidden to ascend to Horus. 586, 645
71 King has come to Horus. 1407, 1489, 1558(E 24)
72 King comes to Nut, having thrown his father to earth and shaken off (*fḫ*) Horus behind him. 250
73 King has come to *Ḥr Šꜣt(y)*, *Ḥr šsmt(y)*, and Eastern Horus treated as one god. 450
74 Actions of king. See also 142(D 168), 337 etc. (D 36), 342(D 35), 360(D 38), 927 etc. (D 37), 999(D 39), 2056(D 165)

b) Position Occupied by Other Divinities in Relation to Horus

1. By others than the king:

75 By *Ḥꜣpy, Dwꜣ-mw·t·f, Ymśty,* and *Ḳbḥ-śn·w·f.* See 601(*C 4*), 2101(D 208)
76 Geb has mystic name "Horus not repeating his work (*Ḥr y·tm wḥm kꜣ·t·f*)." 1622
77 Hostile serpent is asked: "Art thou Horus? . . . Art thou Set?" Apparently negative answer is expected. 685

2. By the king:

IDENTIFICATIONS

78 King is Horus, following context lost. 2147
79 King is Horus who came forth from the Nile (*pr m Ḥʿpy*), the bull that came forth from the walled town (*ng(ꜣw) pr m ynb·t*), the *ḏ·t*-serpent that came forth from Re, the *yʿr·t*-serpent (uraeus) that came forth from Set, say the gods. 2047
80 King is Horus who came forth from the *šnṯ*-serpent. 681
81 King has mystic name "Horus master of his people (*Ḥr ḥry-ḏꜣḏꜣ rḫy·t·f*)." 644
82 King is called "Horus presiding over the spirits (*Ḥr ḫnty yꜣḫ·w*)." 800 (by Re)

[98] Breasted here suggests "bonds"; but in all the other Pyr. occurrences of *šṯ* (1089, 1373, 1507[D 285–87] and ⸢1555⸣) it seems to be used for "adorn" or "adornment."

C. Relationships—

83 "Horus (dwelling) in the sea, Horus presiding over the spirits (*Ḥr ymy w³ḏ wr, Ḥr ḫnty y³ḫ·w*)" are mystic names of Osiris or of king.[99] 1505

84 "Horus presiding over the spirits, star which ferried over the sea (*Ḥr ḫnty y³ḫ·w, śb³ ḏ³ w³ḏ wr*)" are mystic names of Re or of king.[99] 1508

85 King is called "Horus presiding over his ⸢thighs⸣ (*Ḥr ḫnty mn·wt·f*), *Š³ṯwty*, lord of *Śbw·t*." 804(by Re), 1015P(by Anubis)

86 King is called "Horus presiding over his *mny·t*." 1015N(by Anubis), 1719(by Re)

87 King has mystic name "Horus of *Š³·t*." 1257P[82]

88 King is Horus, the young child with his finger in his mouth (*ẖrd nḫn ḏbʿ·f m r³·f*). 664

89 King is Horus the son of Osiris; is the first-born god, son of Hathor; is seed of Geb. 466

90 King is Horus, is 'the son whom he loves (*s³ mr·f*)' of his father Osiris. 1331

91 King is Horus, son of Geb. 973

92 King is Horus who went about in the protection of his eye (*šn m s³-yr·t·f*). 195, 198

93 King has mystic names "Horus of Dewat, Horus-*Śkśn*, Horus (following lost)." 1734

94 King is Horus who adorned his eye (Egypt or some part thereof) with both his arms (*ḏb³ yr·t·f m ʿ·wy·fy tm·wy*). 1596

95 For king as Horus, see also 148(D 180), 316(D 7), 436(G 50), 493(E 4), 503(D 69), 723(D 93), 800(D 51), 852(D 132), 853(D 2), 1086(D 34), 1294(D 194), 1301(D 3), 2036(D 29, C 111), 2037(D 62)

COMPARISONS

96 Isis the Great dries (king) as if he were Horus (*Ḥr yś*). 1140, ⸢[1733]⸣

97 [Children of] Horus [are to carry] king [like (*my*) Horus in (Sokar's) *ḥnw*-barque.] 1823

98 King *ḫśd*'s his waters etc. like (*yś*) Horus, and ⸢distinguishes⸣ (*wp*) them like Wepwawet. 2032

99 King sits before divine ennead like (*yś*) Geb, like Osiris, like Horus lord of men and gods (*nb-pʿ·t nṯr·w*). 895

100 King is like (*yś*) Horus ———.[100] 1915

101 King comes to ⸢Anubis⸣[101] like (*yś*) Horus when he had avenged his father Osiris (*nḏ·n·f yt·f Yśyr*). 573

102 Isis and Nephthys grieve for king as if he were (*yś*) Horus who avenged his father Osiris (*nḏ yt·f Yśyr*).[102] 898

103 King dawns like (*yś*) Horus presiding over the living (*ḫnty ʿnḫ·w*), like Geb, and like Osiris. 2103

[99] Suffix *ṯ* is error for either *k* or *f*.

[100] The *n* following Horus may be initial of *nḏ yt·f*, "avenger of his father," as written in 898aN, or possibly of *nḫn*, "young," as in 1320P.

[101] Cf. 574a.

[102] *Yś* seems to be an error, for context clearly treats king not as Horus but as Osiris.

C. Relationships—

104 King bears Re's speech as if he were the god, as if he were Horus

(when Re says): "I am thy brother, even ⸢(of)⸣ Soped." 1863(cf. C 9)

105 King commands the *ḥnmm·t*-people like (*yś*) Min (dwelling) in his house and like Horus of *Ḏbʿ-ḫrw·t*. 1993

106 For other comparisons of king with Horus, see 4(D 42), 5(D 72), 206(D 274), 346(C 35), 353 etc. (D 44), 362(D 45), 659(D 50), 684(D 14), 768(D 8), 795(G 39), 798 etc. (D 9), 810(D 1), 874(H 6), 881(D 56), 891(D 57), 953(D 232), 1013(G 28), 1089(D 285), 1113(D 19), 1294(D 194), 1373 (D 286), 1507(D 287), 1539(G 18)

MISCELLANEOUS RELATIONS OF THE KING TO HORUS

107 King as Osiris has ascended before (*m bꜣḥ*) Horus. 576

108 King's right side is in Horus and his left side is in Set. 601(*C 4*)

109 King as Osiris has shone (or 'become a spirit') in (or 'through') Horus (*yꜣḫ·n·k ym·f*). 633, 1637

110 King as *Wng*, son of Re, is announced to "these two great and mighty gods" (Re and Horus; cf. 951). 952

111 King has come from (or 'as'?) Horus (*ym·f*). 2036

112 King knows (*rḫ*) the god (*nṯr*), Re, Thoth, *Ḥr špd*, *ymy Dwꜣ·t*, and the bull of the sky (*kꜣ-p·t*); and each of these is bidden to fail not to recognize (*m ḫm*) him. 327–32

113 King is greater than (*ʿꜣ r*) *Ḥr ṯmśty*,[84] the red crown on head of Re (*dšr·t tpy·t Rʿ*). 702

114 King is that single star ascending in east of sky, who has not given his body to Horus of Dewat. 877

115 On miscellaneous relations of the king to Horus, see also 27 etc. (D 267), 133(D 4), 301(D 263), 467(D 224), 535(D 198), 582 etc. (D 192), 856 (D 223), 888(D 5), 1218(D 6), 1254(D 196), 1355(D 269), 1406(D 204), 1690(D 264), 1824(D 104), 1827(D 97), 1988(D 26)

c) *Attitude of Other Divinities toward Horus*

116 Horus is beloved son of Osiris. 179(G 8), 1331(C 90)

117 The watchers (*wrš·w*) purify themselves for Horus. 1945

118 Sky (*p·t*) purifies itself for Re, earth (*tꜣ*) purifies itself for Horus. 951

119 The gods fear before (*nr n*) Horus. 1794

120 On attitude of ⸢Re, Eastern Horus, and Harakhte, as one god⸣, toward Horus, see 1088(D 266)

121 *Rnn-wt·t* loves Horus. 454

122 Horus is beloved son of king as Osiris. 11, 644

123 [King loves] Horus. 45

124 King as Osiris is to show himself gracious to Horus. 103, 611

125 King as Osiris is entreated to accept and be satisfied with the speech of Horus. 611, 646

126 Heart of king is satisfied with Horus. 897

127 King has praised Re, Eastern Horus, and Harakhte. 1087

128 King is to be satisfied with eyes of Horus. 103(F 425), 583(F 426)

D. Nature—

I. Attributes and Powers

a) *Physical*

1 King lives as lives Horus presiding over Letopolis. 810

2 King as "the sole one" (next called Horus, long-striding, mighty over the horizon, mighty over the gods) shall endure always. 853

3 King ascends as (*m*) Horus of Dewat, presider over the Imperishable Stars, and sits upon his marvelous throne at head of his celestial lake (*ḫndw·k byꜣ tp mr·k ḳbḥw*), living as the scarab-beetle lives, enduring as the *ḏd*-column, for ever and ever. 1301

4 King lives on that whereon Horus presiding over estates lives, eats and drinks of latter's provisions; *ḥnḏ* and *šꜣšr·t* are his food. 133

5 King lives on that whereon Horus, lord of the sky, lives, by command of the latter. 888

6 King is to eat and drink of the provisions of Morning-star, Horus of Dewat etc. 1218(E 91)

7 King is Horus, the heir of his father; he is the goer and comer, the fourth of these four gods who fetch water, set down the offering (*ꜥbꜥb·t*), and ⸢do — with⸣ (*yrr·w hyy m*) the thighs of their fathers. 316

8 The going (*šm*) and goings (*šm·wt*) of king are those of Horus. 768

9 King goes (as) Horus goes (*šm*); he speaks (as) Set speaks. 798, 1715

10 The sleepers (*šḏr·w*), the watchers (*wrš·w*), and Horus are all bidden to wake. 1011

11 *Ḥsmnw*, Eastern Horus, Eastern Soul, and Harakhte, treated as one god, wakes in peace. 1478

12 Horus presiding over the spirits, Re, and *Mdy* (=Re?) wake in peace. 1518

13 Horus ⸢collapses⸣ (*sbn*). 503

14 King stretches bow-string as (*pḏ rwḏ m*) Horus, draws ⸢slip-noose⸣ as (*štꜣ wn·t m*) Osiris. 684

15 Horus is able (*sḫm*) himself (alone) to avenge his father Osiris the king. 1334

16 Horus ⸢is not *wnk*⸣. 611, 646

17 Horus and Set are each sound ⸢of body⸣ (*wḏꜣ n ḏ·t·f;* or 'sound because o himself,' i.e., each causing his own soundness). 683

18 Horus is black and great (or 'very black') in his name of *Km wr*. 1658

19 King's sweat (*fd·t*) and odor (*sṯy*) are those of Horus. 1113

20 On physical nature of Horus, see also 7–8(A 14), 25 etc. (E 17), 103 etc. (A 19), 143(B 3), 206(D 23), 253(E 11), 436(G 50), 457(H 4), 663–64(A 77), 695(H 5), 852(G 27), 854(A 17, A 79), 1207(D 61), 1214(G 3), 1268(B 11), 1320(C 49), 1622(C 76), 1807(A 13), 2032(C 98)

b) *Spiritual*

21 See references to Horus as a god (*nṯr*) under A 41–49

22 Horus has become more a soul and more mighty than his father Osiris, and Set than Geb (*bꜣ·n·k yr·f sḫm·n·k yr·f*). 144

23 King as Set suffers not destruction nor [⸢cessation⸣] (*n fḫ·ty n·k n yꜣb·ty n·k*), but is more a soul and more mighty than the Southern gods and their spirits; king as *Ḥr ḥwnty*[68] has same qualities in relation to the Northern gods. 204–6

D. Nature—

24 Horus, lord of the green (cosmetic?), has become a soul and a prepared one (*bꜣ·n·k śpd·n·k*). 457

25 To Horus, heir of Geb, ⸢(so-)called of Atum, belongs everyone whom the two enneads mention and everyone whom he (himself) mentions⸣ (*n·k tm ḏd·w pśḏ·ty n·k tm ḏd·(w)·k;* or possibly 'all which the ... enneads say and all which thou sayest'). 1489

26 Righteousness (*mꜣʿ·t*) of Horus is that of the king. 1988

27 Horus purifies himself in Pe and comes purified [to] avenge [his father]. 2190–91

c) Celestial

28 *Sḫn wr*[103] is filled with splendor (*yꜣḫ*) of *Ḥr Šꜣt(y)*, *Ḥr śsmt(y)*, and Eastern Horus, all considered as one god, as latter ascends in splendor from horizon. 455

29 Re causes king to shine (*ś·(w)bn*) as Horus. 2036

30 On celestial nature of Horus, see also 362(D 45), 585(E 155), 612 etc. (E 157), 633 etc. (C 109), 636(E 156), 702(C 113), 1508(C 84)

II. Habitat

a) Horizon

31 Harakhte ferries over on the two floats of the sky to horizon to Re. 337

32 Re and Harakhte ferry over on the two floats of the sky to horizon to each other's presence. 351, 358, 926, 932

33 Horus ferries over on the two floats of the sky to Re, and Re ferries over in same way to Harakhte. 1103

34 Horus, the Horizon-god (*yꜣḫty*), *Ḥr śsmty*, Eastern Horus, and the king as Horus of the Gods each ferry over on the two floats of the sky to Re to horizon. There king receives his throne (*nś·t*) in Field of Rushes and descends to southern region of Field of Offering. 1084–87

35 *Ḥr* (*śsmty*) and king ferry over on the two floats of the sky to horizon to Harakhte. 342

36 King ferries over on the two floats of the sky to horizon to Harakhte and Re. 337, 351, 358

37 King ascends (*pr*) upon the two floats of the sky to horizon to Re and Harakhte. 927, 933

38 King is ferried ⸢across Lily Lake⸣ to horizon to Harakhte. 359–60

39 King is ferried over to Re and Horus of the Gods to horizon, his station being on east side of sky. 999–1000

40 Horus[104] sails with king to horizon in barque of Re, and together they judge the gods in horizon. 2046

41 King ascends and is hailed by Horus in horizon. 2019

42 King controls (*śḫm m*) horizons like (*yś*) Harakhte. 4

43 Note additional occurrences of Harakhte (Horizon-Horus) under A 3; for other connections with horizon, see also 7b(A 30), 372(E 94), 455 (D 28), 741(G 2), 853(D 2)

[103] Equated with Re in 200–201 and 209.

[104] Equated with Re? Cf. 2045.

D. Nature—

b) Sky

44 King is born in east of sky like (*yś*) Har(akhte). 353, 928, 934

45 Re has [taken] king to east side of sky like (*yś*) Horus of Dewat, the star that illumines (*wpš*) the sky. 362

46 The double doors of sky and of firmament (*ḳbḥw*) are opened at dawn for Horus of the Gods, Harakhte, Eastern Horus, *Ḥr šsmty*, and the king, that each may ascend in Field of Rushes and purify himself in Field of Rushes. 525–29

47 The double doors of sky and of firmament are opened for Horus of the Gods, *Ḥr šsmty*, Eastern Horus, Harakhte, and the king, that each may ascend and purify himself in Field of Rushes. 1408–11

48 The double doors of sky and of firmament are opened for Horus of the Gods, Eastern Horus, *Ḥr šsmty*, Osiris, and the king, that each may ascend at dawn and purify himself in Field of Rushes. 981–85

49 The double doors of sky and of firmament are opened at dawn for Eastern Horus, Horus of Dewat, *Ḥr šsmty*, and the king, that each may descend (*sic*) and purify himself in Field of Rushes. 1132–37

50 The double doors of sky are opened for king to go forth from them like (*yś*) Horus, like the jackal upon his side.[105] 659

51 King ascends to sky as (*m*) Horus who is upon the *šdšd* of the sky. 800

52 Horus upon *šdšd* of sky is entreated to give his arm to king when latter ascends to sky. 1036

53 The ways of the Bows (*wꜣ·wt-pḏ·wt*), which cause (one) to ascend to (i.e., lead up to) Horus, are kept clear for king when he escapes to sky. 801

54 Horus purifies king in the firmament (*ḳbḥw*). 841

55 Horus (here called *bꜣ ymy dšr·f*, 'Soul dwelling in his blood') received his seat in zenith (*wpw·t*) of sky, in the place wherewith his heart was satisfied, that he might course sky and wander over Delta and Upper Egypt (*šꜣšy ˹ydḥ·w˺ šmꜥ*). 854

56 King takes *wrr·t*-crown in sky like (*my*) Horus, son of Atum. 881

57 King has coursed sky as (*m*) Harakhte. 891P[106]

58 House of Horus (*ḥt·t-Ḥr*) is in sky. 1025–27

59 House of Horus is in the firmament (*ḳbḥw*). 1327

60 King traverses (*ḫns*) firmament in wake of Re-Harakhte. 1049

61 Morning-star, Horus of Dewat, divine falcon, ˹bright green˺ (*wꜣḏꜣḏ*, a bird), same divinity in four aspects,[107] is called child of the sky; and his four contented faces see that which is in *Kns·t* and drive away the ˹dew˺ (*ššn*) from the offering-tables. 1207

62 King as Horus ascends to Re, ˹seizing˺ Nut (the sky) by the forelock.[108] 2037

[105] Anubis; cf. 2001.

[106] MN have *snḥm*, "a grasshopper."

[107] Text reads *y·nḏ ḥr·k m fd·w·k ypw ḥr·w ḥtp·w*, "hail to *thee* with these *thy* four contented faces."

[108] This passage reads literally: "N. ascends unto him (Re), Horus ˹seizes˺ (*wrm*) Nut by the forelock (*wp·t* with hair det.)." The verb *wrm* clearly occurs in 524c. Breasted translates *wp·t* as "forelock" in 401a (*Development of Religion and Thought*, 127).

D. Nature—

63 King puts himself on way (*w3·t*) of *Ḥr šsmty*, wherein latter leads the gods to the beautiful ways (*w3·wt*) of the sky and of the Field of Offering. 2062

64 On Horus in sky, see also 7b(A 33), 70(F 427), 304(D 210), 502(E 113), 534(D 135), 802(E 108), 815(E 114), 888(D 5), 999–1000(D 39), 1010 (E 115), 1301(D 3), 1465(E 118), 1979(E 173), 2090(F 199), 2091 (F 119)

c) Field of Rushes (sḫ·t-y3r·w)

65 Horus of the Gods, *Ḥr šsmty*, Osiris, and the king each ascend at dawn after purifying themselves in Field of Rushes. 986–89

66 If Horus of the Gods, *Ḥr šsmty*, Eastern Horus, and Harakhte make an ascent to purify themselves in Field of Rushes, king shall do likewise. 1412–15

67 Re, Horus, and king have purified themselves in Field of Rushes.[109] 1430

68 For connections with Field of Rushes, see also 525–29(D 46), 874(H 6), 981–85(D 48), 1084–87(D 34), 1132–37(D 49), 1408–11(D 47)

d) Way (w3·t) of Horus[110]

69 King passes along a way made by the gods, for he is Horus. 503

70 Way of Horus is bidden to extend its hand (*yr ḫn·ṯ*)[110a] and its two arms (*ᶜ·wy*) to king. 607

71 On way of *Ḥr šsmty*, see 2062(D 63)

e) Dewat

72 King presides in Dewat like (*yš*) Horus, presider over Dewat. 5

73 *Ḥr špd* and *ymy Dw3·t* in vocative have interchanged places in W and T texts of 330 and 331. Are they identified?

74 See also the references to Horus of Dewat under A 96; still other connections appear in 372(E 94), 390(E 111), 1959(F 209)

f) Y3·t-regions

75 King circles (*dbn*) the Horite regions and the Setite regions. 135, 1735

76 [King circles his] Horite [regions] and his Setite regions like Min. 1928

77 King circles the Horite regions; he invades (*dndn*) the Setite regions. 2099

78 King is bidden to go to give command to the regions of Horus, of Set, and of Osiris. 218, 222

79 The region ⌜of Geb⌝, that of Horus, that of Set, and the Field of Rushes praise king. 480

80 The regions of Horus, those of Set, and the Field of Rushes praise king. 994

81 Horus is greeted in the Horite regions, Set in the Setite, and *Y3rw* in the Fields (so W; N has 'Field') of Rushes. 487

82 King belongs to that distant palace of the lords of kas where Re is at morn (*dw3·w*; or 'is praised'), in the Horite regions, in the Setite regions. 598

[109] Cf. also connection of Horus with *Lake* of Rushes in 519 and 1247 (E 95–96).

[110] Cf. "ways of Kheprer" in 305, and "ways of the Bows" in 801(D 53).

[110a] See Lacau, *Rec. de Trav.*, XXXV, 220.

D. Nature—

83 King is to inhabit his Horite regions and wander over (*wnwn*) his Setite regions. 770

84 The Fields of Rushes, the Horite regions, and the regions of Set are all for king. 943

85 The gods who have gone to their kas (*s(y)·w n kȝ·w·śn*) live in the Horite and the Setite regions. 948

86 Sky, earth, Field of Rushes, Horite regions and Setite regions, cities and nomes are given to king by Atum. 961

87 Regions ⌜of Geb⌝, those of Horus, those of Set, and the Field of Rushes are given to king by Atum. 1475

88 Anubis has commanded that king descend as Morning-star and invade (*dndn*) region of southern Horus and region of northern Horus. 1295

g) *Earthly Localities*

89 ⌜The remedies of Horus have been applied⌝ (*pẖr pẖr·wt-Ḥr*)[111] in Abydos (*ȝbḏw*), ⌜(even) the *wy·t*-bread of⌝ Osiris. 1122

90 Horus comes forth from Khemmis (*ȝḫ-by·t*), and Pe (*P*) arises for him. 2190

91 Double Horus (*Ḥr·wy*), presider over estates, is called also Great One (dwelling) in Heliopolis (*Ywnw*). 695(H 5)

92 For Heliopolis, see also 1614(F 72)

93 King's soul stands among the gods as (*m*) Horus (dwelling) within *Yrw*. 723

94 Delta (⌜*Ydḥ·w*⌝). 854(D 55)

95 The Horus hostile to dead is bidden to begone to *ʿnp·t* and to *Nṯr(w)*. 1268

96 Pe (*P*). 2190(D 90, D 27)

97 King [as Osiris] is mighty [over Delta (⌜*Mḥ*⌝) as (over) this Horus.] 1827

98 Hieraconpolis (*Nḫn*). See references in A 39

99 *Nṯrw*. 1268(D 95)

100 Letopolis (*Ḫm*). See references and notes under A 63 and A 70

101 *Ḫȝ·t*. Proper reading of *Šȝ·t* (D 103), *q.v.*

102 King has trodden upon ⌜the Horite city of *Sbn*⌝.[112] 244

103 *Šȝ·t*. 450 ff.(C 73, D 28), 1257(C 87)

104 King as Osiris [is mighty] over Upper Egypt (*Šmʿ*) as (over) this Horus. 1824

105 Upper Egypt. See also 854(D 55)

106 *Kns·t*. 920–21(D 229), 1207(D 61)

107 The Two Lands (*tȝ·wy*). 6–7(A 26), 1258(C 51)

108 King begot (or 'is child of') Horus (dwelling) in *Ḏbʿ-ḫrw·t*, like (*yś*) Set (dwelling) in ⌜*Ḥnḫn·t*⌝. 734

109 *Ḏbʿ-ḫrw·t*. See also 1668(D 190), 1993(C 105)

[111] Breasted renders: "⌜the offering of Horus is offered⌝;" but *pẖr*, "food-offering," is masc. in 818a, the only certain Pyr. occurrence of this root with such meaning.

[112] Breasted: "⌜the *sb* of the city of Horus⌝."

D. Nature—

h) Miscellaneous

110 Eastern (*y³bty*). See references under A 2, also 159(E 121)
111 *Yḫm·wt*. ⸢1951⸣(A 91)
112 The two regions (*ydb·wy*) shall be withheld from Horus,[113] if Sun-god prevents king from coming to place where Sun-god is. 1436
113 King sees those who are in the palace (*ʿḥ*),[47] namely Horus and Set. 141
114 Sea (*w³d wr*). 1505(C 83), 1508(C 84)
115 The broad hall (*wśḫ·t*). 905(E 90)
116 Among the falcons (*byk·w*). ⸢1672⸣(E 164)
117 House or estate (*pr*). 1294(D 194)
118 Morning-barque (*mʿnḏ·t*). 1479(D 124)
119 As Horus does not pass the night behind the lake (*n Ḥr śḏr ḥ³ mr*),[114] nor is Thoth left boatless,[115] so king too is not left boatless. 1429
120 Jackal-lake (⸢*mr*⸣*-s³b*). 372(E 94)
121 Celestial lake (*mr ḳbḥw*). 1301(D 3)
122 "They" (obscure) row Horus at his ascent from (*m*; or 'in'?) *Mḥ·t wr·t*. 1131
123 Northern (*mḥty*). 1295(D 88)
124 *Ḥsmnw*, Eastern Horus, Eastern Soul, and Harakhte, treated as one god, passes the night in evening-barque (*mśkt·t*) and wakes in morning-barque (*mʿnḏ·t*). 1479
125 *Rp·t*-chapel. 767(B 7)
126 Southern (*rśw*). 1295(D 88)
127 *Ḥnw*-barque. 138(B 2), 620(B 5), [1823](C 97), [1824](E 105), ⸢1826⸣ (E 106)
128 The Great House (*ḥt·t ʿ³·t*, probably in sky). 373(E 110)
129 The double doors of the *ś³·t*-region[116] are opened for Horus, and those of ⸢the meadows⸣ (*ś³b·wt*) for Set. 518
130 The star Sothis (*Śpd·t*). 632 and 1636(B 6)
131 Great field (*śḫ·t wr·t*) of Morning-star, Horus of Dewat etc. 1217(E 91)
132 The caverns of the lookouts (*tpḥ·wt-ptr·w*) are opened for king as Horus; the footsteps of radiance are loosed (*śnfḫfḫ nmt·wt-y³ḫw*) for him. 852
133 *Ḏndrw*-barque. 633 and 1637(B 1)
134 The king. 19 etc. (C 45), 44(C 64), 636(C 46), 1633(C 47)

III. ATTITUDE OF HORUS TOWARD OTHER DIVINITIES

135 *Nbyw·t* (a goddess?) in sky is beloved of Horus. 534
136 *Ḥ³py*, *Ymśty*, *Dw³-mw·t·f*, and *Ḳbḥ-śn·w·f* are beloved children of Horus. 1548
137 Horus followed and loved Geb. 1625(G 51)

[113] Breasted interprets that Horus shall be prevented from inheriting Egypt as successor of the Sun-god.

[114] I.e., as Breasted notes, he does not have to remain without crossing it.

[115] These phrases picture Horus as sun and Thoth as moon crossing the sky.

[116] The nisbe form *ś³ty·w* occurs in 1360 and 2017. Cf. also the vocative *ś³ty* in a serpent charm in 421 and 668.

D. Nature—

138 Horus is satisfied with his father (the king); Atum is satisfied with his years. 258

139 Horus is to be satisfied with Osiris the king. 584

140 Horus loves his messenger the king. 535(D 198)

141 Horus has loved Osiris the king. 609

142 Horus has loved his father the king. 1633

143 Horus is not distant from Osiris the king. 610

144 Horus comes rejoicing at approach of king and of his eye which is upon latter. 2076

145 As to attitudes of Horus, see also 592 and 650(D 147), 1088(D 266), 1148(C 69)

IV. PARTS OF BODY

146 Incense is in limbs (⌜*ywᶜ·w*⌝) of Horus. 116

147 Heart (*yb*) of Horus rejoices in presence of Osiris the king. 592, 650

148 Heart (*yb*). See also 71(F 428), 585(D 176), 854(D 55), 1640(G 4)

149 White teeth (*ybḥ·w ḥḏ·w*) of Horus (onions?) are offered to Osiris the king. 35, 79

150 Eye (*yr·t*) and eyes (*yr·ty*). See section F as a whole.

151 A hostile bull's head is in hand (ᶜ) of Horus, his tail in hand of Isis, and Atum's fingers are on his horns (*wp·t*). 1302

152 Arm (ᶜ) of Horus is behind king, and arm of Thoth —. 1570

153 Horus has caused that his arm ⌜revert to⌝ (*pḫr ᶜ·f n*) [king]. 2213

154 Arm and arms. See also 741(G 2), 1036(D 52), 1243(E 109), 1588 etc. (F 91)

155 Nose (*fnḏ*) ⌜of (*Ḥr*) *šsmt(y)*⌝ (or of Sobk?) breathes perfumes. 456

156 *Pḳ*-bread and *pꜣd*-bread came forth from loins (*mꜣś·t*) of Horus. 378

157 Thighs (*mn·wt*). ⌜804 etc.⌝(A 68)

158 Tip of breast (*tp n mnḏ*) of Horus, of his body, is offered to king. 32

159 Breast of Horus is offered to Osiris the king. 91

160 Horus and Set respectively were purified and healed from the wounds they had given each other by the spittle (*yšš*) which came forth from the mouth (*rꜣ*) of Horus and by the spit (*yšd*) which came forth from the mouth of Set. 850

161 Mouth. See also 663–64(A 77)

162 Arm (⌜*rmn*⌝) of Horus fights in behalf of *šḥd·w* of sky, whom Re causes to live (again) every day. 449

163 King descends . . . on shoulders (*rmn·wy*) of Horus. 138

164 Horus has set king (upon) his shoulders. 1471

165 King sits at shoulder of Horus and spits on his temple (*śmꜣ*).[117] 2056

166 Ointment which was in forehead (*ḥꜣ·t*) of Horus is put in forehead of king. 52

167 Forehead. See also 83(F 246), 139(F 121), 453(F 38), 742(F 39, F 158, F 217)

[117] King does same for Osiris in 2055.

D. Nature—

168 King spits on face (*ḥr*) of Horus to expel the ⸢wickedness⸣ which is upon him (*nkn yr·f*), and *yꜥḥ*'s testicle of Set to expel his 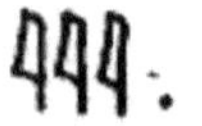 142

169 *ꜣkr* and Horus each bare the other's face. 555

170 Face. See also 973(D 234), 1207(D 61)

171 A serpent is called *ḫꜣnf·t* of Horus (with flesh det. in W). 245

172 Neck (*ḫḫ*). 1213(D 181)

173 Body (*ẖ·t*). 647(C 24)

174 Temple of the head (*šmꜣ*). 2056(D 165)

175 ⸢"Talons"⸣ (*šmꜥ·wy;* or props of some sort?) of Horus and wings of Thoth are bidden to ferry over the king. 1176

176 Bones (*ḳrš·w*) of Horus *nwꜣwꜣ*[118] and his heart (*yb*) beats. 585

177 Foot (*ṯbw*). 681(G 42)

178 Blood (*dšr*). 854(D 55)

179 Body (*ḏ·t*). 32(D 158)

180 King's head (*ḏꜣḏꜣ*) is that of Horus of Dewat, his forehead (*m·ḫnt*) is that of *Ḫnty-⸢yr·ty⸣*, his two ears and again his two eyes are the two daughters of Atum, his nose (*fnḏ*) is that of the jackal, his teeth are Soped, his arm(s) are *Ḥꜣpy* and *Dwꜣ-mw·t·f*, his legs are *Ymšty* and *Ḳbḥ-šn·w·f*, and his members (*ꜥ·wt*) are the two daughters of Atum. 148–49

181 The two children of Nut (namely Morning-star, Horus of Dewat etc. and the king) embark on the sea, each wearing on head (⸢*ḏꜣḏꜣ*⸣) and neck (*ḫḫ*) garlands (*wꜣḥ·w*) of the *yb*-tree. 1213

182 Head. See also 1239(F 197)

183 King has turned (*šṯp*) to sky at the two fingers (*ḏbꜥ·wy*) of the god, the lord of the ladder (Horus, cf. 973–74). 980

184 King's mouth is opened by Horus with his little finger (*ḏbꜥ nḏš*). 1330(G 9)

185 Finger and fingers. See also 372(E 94), 663–64(A 77), 1208(G 32)

V. ELEMENTS OR PHASES OF PERSONALITY

186 Spirit or spirithood (*yꜣḫ*). 633 etc. (B 1), 795(G 39)

187 Soul of Horus is not repelled (*n ḫšf bꜣ·f*). 253

188 Horus becomes a soul. 580, 767

189 Soul. See also 144(D 22), 206(D 23), 457(D 24), 854(D 55), 1209(G 32)

190 Name (*rn*) of Horus endures in *Ḏbꜥ-ḥrw·t*. 1668

191 Horus, Set, Thoth, *Spꜣ*, Osiris, *Ḫnty-⸢yr·ty⸣*, and the king go (*sy*), each with his ka (*ḥnꜥ kꜣ·f*). 17

192 King as Osiris is ka of Horus. 582(E 22), 610, 647(E 22), 1609, [1832]

193 Eye of Horus was ⸢before⸣ (*ḥry·t ꜥ·wy*) his great kas and upon (*tpy·t*) his many kas. 2087

VI. SUBORDINATES

194 Horus of Hieraconpolis has given to king as to *Ḥr ymy pr·f*, as to the presider presiding over the mighty, his spirits the jackals (*yꜣḫ·w·f sꜣb·w*). 1294

118 Breasted: "dance."

D. Nature—

195 Spirits. See also 800 etc. (A 64), 969N(E 140)
196 King comes as messenger (*yp(w)t(y)*) of Horus. 1254
197 ⌜Priest[119] has come to Osiris the king as messenger of ⌜Horus⌝ (*yp(w)t(y)-Ḥrw*). 1686
198 King is beloved messenger (*ynw*) of Horus, Set, and Thoth, for he brought (so T; N has 'brings') to them their eye, testicles, and arm (last in N only) respectively. 535
199 Messengers of Horus go, his couriers run (*sy yn·w·f bṯ šyn·(w)·f*) to announce coming of king to Re.[120] 253–54
200 Messengers of Horus go, his couriers run, his heralds hasten (*nṯꜣ y·ḥw(w) t(y)·w·f*)[121] to announce coming of king to Re. 1861–62
201 Doorkeeper of Horus, [ga]te-[⌜keeper⌝] of Osiris (*yr(y)-ʿꜣ n Ḥr* [⌜*yry*⌝-*ʿrr*]*w·t n·t Yśyr*) is bidden to announce king to Horus. 520
202 *Yḫm·wt.* ⌜1951⌝(A 91)
203 Isis. 584(E 62)
204 King judges (*wḏʿ mdw n*) the gods like (*yś*) a king, like the deputy (*yśty*; or 'successor') of Horus, that he may avenge Horus's father Osiris (N has: 'who avenged his father Osiris'). 1406
205 Horus stands before the living (*ḫnty ʿnḫ·w*). 1232
206 The living. See also 2103(C 103)
207 *Wr-kꜣ·f* ('Great is his ka'), the butler (*wdpw*) of Horus, mighty in counsel (*šḫm-sḥ*) of Re, eldest of the palace (*śmśw-ys·t*) of Ptah, is bidden to give to king a joint of meat (*wr*). 560, 566
208 Horus comes to king, equipped with [his] souls (*bꜣ·w*), namely *Ḥꜣpy*, *Dwꜣ-mw·t·f*, *Ymśty*, and *Ḳbḥ-śn·w·f*, who bring to king his name of 'Imperishable.' 2101–2
209 People (*pʿ·t*). 14 etc. (A 34), 895(C 99)
210 There is conflict in the sky . . . and the divine ennead of Horus (*pśḏ·t-Ḥr*) is in splendor (*m yꜣḫw*) . . . as king sits in every seat of Atum and takes the sky. 304–5
211 Divine ennead. See also 895(C 99)
212 Children of Horus. 24 etc. (E 137), 619(E 101), 637 etc. (E 102), 1828 (E 138)
213 Nephthys. 584(E 62)
214 Cattle of Horus, whom he treads (*nr·w-Ḥr ḫnd·(w)·f*), may apply to *ḏ·t*-serpent and *spꜣ*-worm mentioned just previously. 244
215 Gods (*nṯr·w*). 24 etc. (E 125–36), 577 etc. (E 38, E 161), 578(E 61), 895(C 99), 1458(E 123)
216 The sweet-smelling ones (*nḏm·w-śt(y)*). 1518(A 92)
217 People (*rḫy·t*). 644(C 81)
218 Heralds (*ḥwwty·w*). 769(D 221), 1861(D 200)
219 People (*ḥnmm·t*). 1993(C 105)
220 The Imperishable Stars (*y·ḫm·w śk*). 1301(D 3)

[119] Horus himself speaks in 1683–85, but throughout 1686 *Ḥrw* is in 3rd person.

[120] Here unnamed, but cf. 1861–62(D 200) and 2174–75.

[121] Same grouping, but with different verbs, is used of king's envoys in 1539 (following G 18).

D. Nature—

221 Couriers of Horus run, his heralds hasten (*bṯ šyn·w·f nṯꜣ ḥwwty·w·f*) to announce him to him who leans (upon his staff) in the East (Re).[120] 769

222 Couriers (*šyn·w*). See also 253(D 199), 1861(D 200)

223 He who knows chapter of Re and recites charms of Harakhte (conditions met by king) shall be intimate (*rḫ*) of Re and companion (*śmr*) of Harakhte. 855–56

224 Osiris has commanded that king dawn as deputy (or 'successor') of Horus (*śn·nw-Ḥr*). 467

225 The *śḫꜣ·t-Ḥr* (cow) suckled king. 1375

226 The mighty (*śḫm·w*). 1294(D 194)

227 King has been presented with *ḥsmn*-natron along with the Followers of Horus (*šmś·w-Ḥr*). 26

228 King satisfies the followers of Horus. 897

229 King descends to Field of *Kns·t*, that the Followers of Horus may purify him. They cleanse and dry the king, and recite for him the 'chapter of the just' and the 'chapter of them that have ascended' (P adds: 'to life and satisfaction'). 920–21

230 King is purified by the Followers of Horus, who also recite for him the 'chapter of them that have ascended' and 'them that lift themselves' (*św·yw* in P; but MN have *rꜣ n ḫp·yw*, 'the chapter of them that go'). 1245

231 (The inhabitants of) the Two Lands (*tꜣ·wy*). 1258(C 51)

232 They of Dewat (*D(w)ꜣty·w*) are bidden to support king like (*my*) Re, follow him like Horus, exalt him like Wepwawet, and love him like Min. 953

233 Thoth. 575(E 56), 651(E 57), 1336(E 59), 1979(E 60)

VII. Equipment

This covers in one alphabetic series:

Objects worn or carried
Food, cosmetics, natron, incense
Palaces, thrones, estates, temples
Barques

Miscellaneous, including:

pẖr·wt
mꜣḳ·t, 'ladder'
ḥꜣ·t, 'tomb'
ḥkꜣ·w, 'charms'
ḥtm, 'be equipped,' in general statement
ḫnsw·t
św·t
śmꜥ·wy

234 Horus comes with his lion-mask upon his head (*ꜣ·t·f tp·f*); his face approaches his father Geb. 973

D. Nature—

235 *Db·t-nḥḥ·wt* has [come] to king ⸢after latter carried off⸣ (*sḫs·n·f*) one of the two *ꜣmś*-scepters of Horus[122] which were in her charge. 522
236 Horizon offers incense (*ydy*) to Horus of Hieraconpolis. 295, 296
237 Palace (*ꜥḥ*). 141(D 113)
238 Garlands (*wꜣḥ·w*). 1213(D 181)
239 Green cosmetic (*wꜣḏ*). 457(D 24)
240 Sorceress (*wr·t-ḥkꜣ·w*, a crown). 823(C 56)
241 Southern crown (*wrr·t*). 14(E 86), 455(E 1), 737(F 149), 881(D 56)
242 The broad hall (*wśḫ·t*). 905(E 90)
243 Bread (*pꜣḏ*). 378(D 156)
244 ⸢Pellets of incense⸣ (*pꜣḏ·wy;* or loaves?). 905(E 90)
245 House or estate (*pr*). 1294(A 10).
246 Estates (*pr·w*). 133(A 66), 695(H 5)
247 ⸢Remedy⸣ (*pẖr·t*) and ⸢remedies⸣. 1088(E 169), 1122(D 89)
248 Bread (*pḳ*). 378(D 156)
249 Re dawns against, and Horus draws his nine bows (*pḏ·wt*) against, this spirit which came forth from the earth, ... *Ddy*, son of *Śrḳ·t-*⸢*ḥtw*⸣. 673
250 Staff (*mꜣw·t*). 1212(D 255)
251 Ladder (*mꜣḳ·t*) of Set and of Horus was made for Osiris, that he might ascend thereon to sky and join court of Re (*śtp·f sꜣ yr Rꜥ*). 971
252 Geb gave to Horus (or Osiris? Cf. 973b vs. 971e) the ladder of the god (Horus, cf. 973) and of Set, that he might ascend thereon to sky and join court of Re, and is now entreated to do same for king. 974–75
253 All spirits and gods who shall oppose king when he ascends to sky upon ladder of the god (Horus, as before) are cursed, and those who shall aid him are blessed. 978–80
254 The god (Horus) is called lord of the ladder. 974 etc. (A 36)
255 Morning-star, Horus of Dewat etc. receives his favorite *mꜥbꜣ*-harpoon, his staff (*mꜣw·t*) which *ḫmꜥ*'s the canals, whose twin barbs (*bwn·wy*) are the rays of the sun, whose twin bone(-point)s (*ḳrś·wy*) are the claws of *Mꜣfd·t*. With this his companion the king is to cut off the heads of the ⸢adversaries⸣ (*dꜣyty·w*) who are in the Field of Offering. 1212
256 Sustenance of truth (*m·ꜥnḫ·t n·t mꜣꜥ·t*). 815(A 61)
257 Morning-barque (*mꜥnḏ·t*) 1479(D 124)
258 Horus has ointment (*mrḥ·t n Ḥr*), Set has ointment. 2071
259 Ointment (*mrḥ·t*). See also 52(D 166)
260 Evening-barque (*mśkt·t*). 1479(D 124)
261 Horus comes filled with ointment (*m[ḏ·t]*); he has embraced his father Osiris. 1799
262 Horus has filled himself with ointment (*mḏ·t*). 2072
263 King has inherited Geb and Atum; he is upon throne (*nś·t*) of Horus the first-born (*śmśw*) (of the gods). 301

[122] The *ꜣmś*-scepter is frequently mentioned. In 43–45, 47, and 731 also the Horus-bird appears, but seems in those cases a part of the phonetic writing of *ꜣmś*. Erman, however (*Gram.*³, §186), reads there *ḥr·ś* as a compound noun formed with the feminine pronominal ending, parallel to *šmꜥ·ś* and *mḥ·ś*, old names for crowns of South and North respectively.

D. Nature—

264 These two great and powerful gods who preside over the Field of Rushes have placed king upon throne of ⸢Horus⸣ (*Ḥrw*). 1690

265 Throne (*nś·t*). See also 1086(D 34)

266 ⸢Re, Eastern Horus, and Harakhte, treated as one god,⸣ is to be gracious to king as he is to Horus upon his throne (*nṯ*) and vice versa. 1088

267 King's *nṯry*-natron is that of Horus, of Set, of Thoth, and of *Spꜣ*. 27, 28

268 *Rp·t*-chapel. 767(B 7)

269 The four goings of king, raised by Geb (1353), are before the tomb (*ḥꜣ·t*) of Horus, whereby one goes to the god when the sun goes down. 1355

270 Prime oil (*ḥꜣt(y)·t*). 742(F 39)

271 *Ḥnw*-barque. 138 etc. (D 127)

272 Charms (*ḥkꜣ·w*) of Harakhte. 855–56(D 223)

273 House (*ḥt·t*). 1025–27(D 58), 1327(D 59)

274 King is equipped (*ḥtm*) as Set and as *Ḥr ḫwnty*. 205–6

275 ⸢(*Ḥr*) *šsmty*⸣ (or Sobk?) causes ka of king to ascend to latter's side as former's *ḫnsw·t* ascends to him. 456

276 Horus sits upon this his marvellous throne (*ḫnd·f pw byꜣ*). 2091

277 Marvellous throne. See also 1301(D 3)

278 *Smyn*-natron. 26(E 88)

279 Arbor (*sḥ*). 2100(E 175)

280 Horus is equipped with his *św·t*-⸢plant⸣. 2072

281 ⸢⸢Props⸣⸣ (*śmꜥ·wy*). 1176(D 175)

282 Morning-star, Horus of Dewat etc. is a soul dawning in the bow (*ḥnt*) of this his ship (*śmḥ*) of 770 cubits (in length), which the gods of Pe bound (*śp*) for him and the Eastern gods bent (*ꜥrḳ*) for him. He is asked to take king with him in its cabin or hold (*šnꜥw*). 1209

283 Incense (*śnṯr*). 116(D 146), ⸢452⸣(B 10)

284 Horus, Set, Thoth, *Spꜣ*, and the king, each one has adorned himself with that (so M; N: 'his') *šsm·t* which traversed (*nmt·t*) the face of his land. 1612–14

285 King has gone forth from Pe unto the souls of Pe, adorned (*šṯ*) with the adornment of Horus, clothed (*ḥbś*) in the raiment of Thoth. 1089

286 King has gone forth from Pe adorned as (*šṯ m*) Horus, bedecked as (*dbꜣ m*) the two divine enneads. 1373

287 King as a Heliopolitan goes forth from his house (*pr*) adorned as Horus, bedecked as Thoth. 1507

288 Regalia (*šṯ*, same word as preceding). See also 1285(C 66)

289 King hungers not, having eaten the *ḳmḫw*-bread[123] of Horus, made for latter by his eldest woman(-⸢attendant⸣, *s·t·f wr·t*) that he might be sated thereby and take this land (Egypt. So T; but MN have *sp·f*, 'his ⸢virtue⸣') thereby. 551

290 Linen (*tꜣyt(y)·t*). 2094(E 174)

291 Sandal (*ṯb·t*) of Horus, in a serpent-charm with obscure context. 444

[123] On affinity of *kmḫ* to Semitic קמח see Bondi (*Lehnwörter*, p. 77) and Burchardt (*Altkan. Fremdworte*, No. 984, and I, §123).

D. Nature—

292 Sandal (*ṯbw·t*) of Horus trampled (*ḥnd*) the *nḥy*-serpent, the *nḥy*-serpent of Horus the young child with his finger in his mouth. 663

293 Horus the son of Isis journeyed over the land in his two white sandals (*ṯbw·ty*), going to see his father Osiris. 1215(G 3)

294 Food (*ḏfꜣ·w*). 695(H 5)

295 *Ḏndrw*-barque. 633 and 1637(B 1)

E. Activities of Horus—

I. INVOLVING DIVINITIES OTHER THAN KING

1 *Ḥr Šꜣt(y)*, *Ḥr šsmt(y)*, and Eastern Horus, treated as one god, is bidden to take *wrr·t*-crown from the great and powerful ʳ—ˡs (*ꜣʿʿ·w wr·w ʿꜣ·w*) who preside over Libya. 455

2 *ꜣkr*. 555(D 169)

3 The king and Morning-star, Horus of Dewat etc. cause to flourish (*s·ꜣḫꜣḫ*) the *n·t*-crowns of the Field of Offering for Isis the Great. 1214

4 King is Horus, has come behind (*yy m ḫt*) his father Osiris. 493

5 King as the son Horus seeks to see (*dbḥ mꜣ·f*) Osiris. 1129

6 On activities of Horus involving Osiris, see also 1643(F 56), and cf. references to king in E 16 ff.

7 ʳHorusˡ traverses ʳNephthysˡ (*šꜣš·f Nb·ʳ(t)ˡ-ḥt·t*,[124] in a serpent charm). 444

8 The Two Lands beam when Horus bares the face(s) of the gods (*y·wn·ȷ ḥr-nṯr·w*).[125] 373

9 The gods. See also 24 etc. (E 125–36), 51(F 28), 502(E 113), 969 etc. (E 122–23), 1406(D 204), 2046(D 40)

10 Re. 337 etc. (D 31–34), 449(D 162)

[124] Possibly compare

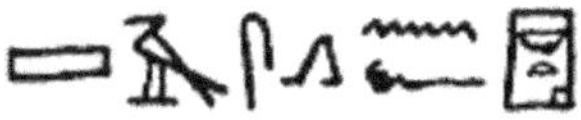

said in 1278 of him who shall favor king's pyramid. On the other hand, text of 444 reads more fully

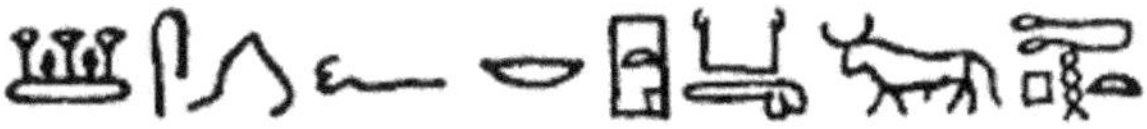

. . . , which suggests also

in 189a (the animal in latter group is a variant of

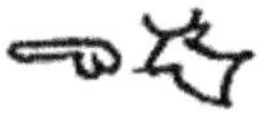

in 547a). Name of Nephthys is not written in exactly this way elsewhere in Pyr.

[125] Sense seems to be "when he appears to the gods," sun rising in presence of the gods and shining upon Egypt; but in Coptic such cases require a dative *n*. Cf., e.g., Zoega, *Catalogus*, p. 305:

ⲈⲢϢⲀⲚ-ⲞⲨⲀⲄⲄⲈⲖⲞⲤ ⲞⲨⲰⲚϨ ⲚⲀⲔ ⲈⲂⲞⲖ.

E. Activities of Horus—

11 Blue-eyed Horus comes up to (*yy r*) and red-eyed Horus protects (*sꜣ*) two wild bulls (*šmꜣ·wy*), when he of the lion-mask[126] is sick (*mr ꜣt(y)*). 253
12 Set. 65(F 90), 95(F 83), 591(F 51)
13 Geb. 1643(F 48)
14 Horus has cast headlong (*gbgb*) the serpent *Ddy*, so that he lives not; Set has *ynyn*'ed him, so that he stands not. 678
15 *Ddy*. See also 673(D 249)

II. Involving King

a) *Favorable*

RECOGNITION

16 Horus recognizes (*yp*) Osiris the king. 587, 612
17 Youthful Horus (*Ḥr rnpy*) recognizes Osiris the king. 25, 767
18 Horus recognizes his father in Osiris the king. 580, 589, 767
19 *Ḥr špd* knows king. 330–31(C 112)

RECITATION

20 Recitation which Horus made for his father the king. 257
21 Horus commanded to make (a recitation) for his father, the cloud-lord Orion (*wḏ·n·f yr·t n yt·f . . .*; might also possibly be 'he assigned the eye to his father . . .,' Orion here representing the king). 261
22 Horus has made (a recitation) for his ka, even Osiris the king, that latter may be satisfied. 582, 647
23 Recitation by Horus. 583(Osirian), 1264(non-Osirian)
24 King has come to Horus that latter may recite for him the powerful and good saying (*mdw ꜥꜣ nfr*) which he gave to Osiris, that king may thereby become powerful and great (*ꜥꜣ wr*). 1558
25 The son Horus has come to speak in behalf of Osiris the king. 1813

RESURRECTION OR RESUSCITATION

26 Horus has caused Osiris the king to stand. 617, 640
27 The son Horus bids Osiris the king to stand. 1683
28 Horus and Thoth raise (*ṯs*) Osiris (the king) (from) upon his side and cause him to stand among (*m* in PM; but N has *m ḫnt*, 'before') the two divine enneads. 956
29 King is bidden to give his hand (*ꜥ*) to Horus, that latter may cause him to stand. 1627, cf. 1796
30 Horus bids Osiris the king come forth (from tomb?) and awake.[127] 1753
31 Horus comes to king, parts his bandages, and casts off his bonds (*wḏꜥ·f sꜣr·w·k ḫꜣꜥ·f mḏ·wt·k*). 2202

RESTORATION OF BODY

32 Libation is poured by Horus for Osiris the king.[128] 24, 765
33 Libations. See also 22(G 10)

[126] Geb; cf. 1032c.

[127] Unnatural order of events.

[128] Both incense and libation were intended to restore to the dead body its lost moisture. See A. M. Blackman in *Zeitschrift für äg. Sprache*, L, 69.

E. Activities of Horus—

34 Horus has given to Osiris the king the latter's waters (*mw*). 106
35 Horus has come to unite (*yʿb*) his father Osiris the king. 101
36 Horus has united Osiris the king. 584
37 Horus has united Osiris the king's limbs (*ʿ·wt*). 617, 635
38 Horus has caused that the gods put Osiris the king together (*dmḏ*).[129] 577, 645
39 Horus has put Osiris the king together, so that there is no disorder (*ẖnnty*) in him. 617, 635
40 The son Horus relieves intestinal pain which king suffers from having eaten an eye, so that king may live thereby. 192
41 Horus forms (*ḳd*) king and causes him to live (again) every day (as does Re the *šḥd·w* of the sky in 449). 450
42 Horus has caused Osiris the king to live. 614, 646
43 Horus has opened (*wp*) eye of Osiris the king, that latter may see therewith. 610, 643, 1806
44 Horus established (*šnṯ(y)*) for Osiris the king latter's two eyes. 644
45 Horus has not permitted that Osiris the king perish (*snw*). 617, 637
46 Horus does not permit that Osiris the king perish. 1753
47 Horus has not permitted that Osiris the king's face *nḫrḫr*. 644
48 On bodily restoration, cf. also 905(E 90), 1684(E 99)

PUNISHMENT OF ENEMIES

49 (When) Horus has spoken, Set bows himself (*yšd*) and bears (*wṯs*) king. 261
50 Horus has seized Set and put him beneath Osiris the king to bear (*wṯs*) latter and tremble beneath him as earth trembles. 581
51 Horus has put Osiris the king's enemy (*ḫfty*) beneath king's feet. 637
52 Horus has put Osiris the king's enemy beneath him to bear him. 642
53 Horus has caused Osiris the king's enemy to bear him. 649
54 Horus has put Osiris the king upon back of his enemy, that latter may not attack (*ẖ3yw*) king. 651
55 Horus brings Set to king, has given Set to him, bowed down (*kšy*) beneath him. 1632
56 Horus has caused that Thoth drive back (*š·ḫt*) for Osiris the king Set's followers, and has brought them to king united (*ywn·w*), after repulsing (*š·ḥm*) for him the heart of Set. 575–76
57 Horus has caused that Thoth bring to Osiris the king latter's enemy. 651
58 Horus comes to reclaim Osiris the king from the gods (*yp·f ṯw mʿ nṯr·w*). 609
59 Horus has come to reclaim his father Osiris the king, and has proclaimed a royal decree (from) upon the seats of Anubis, bidding Thoth spare not those who wronged the king. 1335–36
60 The son Horus has given to king the gods, his enemies; Thoth brings them to king. 1979
61 Horus has caused that the gods avenge (*nḏ*) Osiris the king. 578
62 Horus has caused that Isis and Nephthys avenge Osiris the king. 584

[129] *Dmḏ* is used of putting together members (*ʿ·wt*) in, e.g., 318, 828, 835. In the last two cases *dmḏ* is parallel to *yʿb*.

E. **Activities of Horus**—

63 Horus avenges king. 897

64 Horus avenges Osiris the king. 633, 634, 636, 1753

65 Horus has avenged king. 898

66 Horus has avenged Osiris the king. 582, 649, 1609, 1832

67 Horus has avenged Osiris the king without delay. 591, 618, cf. 1797

68 Horus has avenged his father the king. 1633, 1637

69 Horus has avenged his father (the king). 1728

70 Horus avenges what Set did against Osiris the king. 592

71 Horus has avenged king on his enemies. 758

72 The son Horus has smitten enemy of Osiris the king. 578

73 Horus has smitten and repelled for Osiris the king his enemy Set. 587

74 Horus, avenger of his father, has smitten him who smote Osiris the king, has avenged king on him who did evil against him. 1685

75 The son Horus smites him who smote Osiris the king, binds him who bound him, and puts enemy under his (Horus's)[120] eldest daughter, (dwelling) in *Ḳdm*. 1007–8PM

76 The son Horus has smitten him who smote Osiris the king, has slain him who slew him, has bound him who bound him, and has put enemy under king's eldest daughter, (dwelling) in *Ḳdm*. 1976–77

77 Horus has caused that Osiris the king seize his enemies so that none of them escape (*pśḏw*) him. 579

78 Horus has caused that Osiris the king recognize (*yp*) Set . . . and seize him with his hand (*ḏr·t*) so that he escape not (*m nhp·f*) from him. 582

79 Horus has rescued (*nḥm*) Osiris the king from his enemy. 649

80 Horus has stretched out (*pḏ*) Osiris the king's enemy beneath king. 650

81 Horus has hacked off thighs of Osiris the king's enemies (*śtp·n Ḥr ḫpš·w-ḫfty·w·k*) and brings them to king cut up, after removing their ka(s) from them. 653

82 Horus has repelled king's enemy; the *ꜣkr·w* seize him not. 2202

83 On punishment of king's enemies through Horus, see also 1212(D 255), 1264–65(E 180), 1285(C 66), 1334(D 15)

MOUTH

84 Horus has pressed (*mḏr*) Osiris the king's mouth. 11, 644

85 Horus has opened (*wp*) Osiris the king's mouth. 11, 589, 644

86 Horus opens (*wn* and [*wp*]) king's mouth with that wherewith he opened (*wp*) mouth of his father Osiris, with the metal (*byꜣ*) which came forth from Set, the adze [of metal which opened the mouth(s) of the gods, that king may go and speak in person (*ḏ·t·f*) before the divine ennead in Prince-house in] Heliopolis and take the *wrr·t*-crown from (*ḫr*) Horus, lord of men (*nb-pꜥ·t*). 13–14

87 King's mouth is opened (*wp*) by Horus with his little finger. 1330(G 9)

88 Horus, Set, and the two *tt·wy-yb* spew out *smyn*-natron which opens (*wp*) king's mouth. 26

89 Horus has balanced Osiris the king's mouth against his bones. [1]2, [13], 644

120 But N, as in 1977(E 76), has "thy," referring to king.

E. Activities of Horus—

FOOD

90 ⌜Son of dead king⌝ as Horus[131] gives to king a royal offering of bread and beer and the two ⌜pellets of incense⌝ (*pꜣḏ·wy;* or loaves of bread?) which came forth from Horus dwelling in the broad hall (*ymy wśḫ·t*), that he might satisfy (*ś·ḥtp*) king's heart therewith. 905N

91 Morning-star, Horus of Dewat etc. is asked to take king with him to this his great field (*śḫ·t·k tw wr·t*), which he made to fall ⌜because of⌝ the gods (*m ḏr nṯr·w*),[132] of the evening and morning (meals) of which he eats, which is filled with ⌜Food⌝(*Ḥw*), that king may eat and drink of the god's provisions. 1217–18

92 Morning-star, Horus of Dewat etc. is to place king's satiety above that of *Yw·t*, the presider over her sisters. 1218

93 On furnishing of food by Horus, see also 695–96(H 5), 888(D 5)

PURIFICATION

94 Horus receives king to his two fingers, purifies (*ś·(w)ʿb*) him in the jackal-lake (⌜*mr*⌝-*śꜣb*), cleanses (*ś·fḫw*) his ka in the Dewat-lake, and purifies (*śk*) the flesh of his bodily ka with this which is upon Re's shoulders in the horizon. 372

95 (After) king and Re have purified themselves in Lake of Rushes, Horus wipes king's flesh and Thoth his legs. 519

96 (After) king has purified himself in Lake of Rushes, where Re did the same, Horus is to wipe his back, and Thoth his legs. 1247

97 Horus has expelled the evil which was upon king in latter's four-day period; Set has known not what he did against king in latter's eight-day period. 746

98 Horus of Hieraconpolis has given to king the four *nmś·t*-jars and the four *ʿꜣb·t*-jars . . . wherewith king is purified. 1293

99 Horus has come to purify and cleanse (*wʿb ś·(w)ʿb*) Osiris the king, to revivify (*ś·ʿnḫ*) him, to collect (*ynḳ*) for him his bones, to bring together (*śꜣḳ*) for him his *nb·t*,[133] to collect (*ynḳ*) for him his knives (*dmꜣ·wt*). 1684

100 On purification, see also 841(D 54), 1978(C 54)

CARRYING AND BEARING

101 Horus has given to Osiris the king his children to bear (*wṯs*) king. 619

102 Horus has given to Osiris the king his children to carry (*fꜣ*) king. 637, 1829

103 Horus has carried Osiris the king and bears him. 620

104 Horus carries king, Set lifts (*ś·ṯs*) him. 1148

105 [Horus has carried Osiris the king in *ḥnw*-bar]que; he bears him [as] he bore his father. 1824

106 [Horus] has carried [Osiris] the king; [he bears him in] ⌜*ḥnw*-barque⌝. 1826

[131] P has  .

[132] Breasted suggests "since (the time of) the gods."

[133] Det. is a man swimming.

E. Activities of Horus—

ASCENT[134]

107 Horus asks Osiris the king to give him his hand (ʿ) ———. 68
108 (After) king has traversed Lily Lake in north of Nut (the sky), (Horus) of Dewat takes his hand and leads him (*šḳr D(w)ȝt(y) d·t·k*) to place where Orion is. 802
109 Horus is asked to give his arm (ʿ) to king and to take (*m*) to himself his eye, which seems to be in king's head, that eye and king may both ascend to him. 1243
110 Horus brings king's bodily ka to the Great House (*ḥt·t ʿȝ·t*). 373
111 Horus and Set help king ascend to Dewat upon a ladder made for him by Re. 390
112 Re and Horus stand on either side of the ladder and ⌜lift⌝[135] it before (*ḫft*) Horus's father Osiris (here the king) at his going to his spirit. 472
113 ⌜[Horus] is asked to open for king⌝ the double doors of sky with the flame that is under the ⌜"kettle"⌝ of the gods (*ḥr bḫḫw ẖr ykn·t-nṯr·w*). 502
114 Horus, master of his sustenance of truth, is bidden to seal the double doors of sky and keep back the approachers of its doors as soon as he has taken king's ka to sky. 815
115 The son Horus leads Osiris the king in ways of sky. 1010
116 Horus is asked to take king with him, not to leave him boatless. 1030
117 Horus is asked to take king with him, and Thoth to ferry king over on tip of his wing. 1429
118 Horus opens (*wn*), Set protects (*ḫw;* or '(tries to) prevent'?) that king may shine in east side of sky like Re. 1465
119 King as son of Re is caused to come to him by Horus, Set, Geb, and the souls of Heliopolis and of Pe. 1492–95
120 For other connections of Horus with king's ascent, see 69–70(F 427), 456(D 275), 980(D 183), 1036(D 52), 1176(D 175), 1208–9(G 32, D 282), 1213(D 181), 1471(D 164), 1570(D 152), 2019(D 41), 2046(D 40), 2106(E 150)

DEIFICATION

121 Set and Nephthys, Osiris and Isis, Thoth, and Horus are bidden to go and announce to the southern, northern, western, and eastern gods (except last group, called 'souls') respectively and their spirits the coming of the king as an imperishable spirit. 153–59
122 Horus, son of Osiris, puts king among the gods. 969P
123 King associates with the gods of the Underworld (*nṯr·w* ⌜*nty·w*⌝), the Imperishable (Stars), leaning with them upon a *wȝś*-scepter and a *ḏʿm*-scepter, by command of Horus, the hereditary prince, the king of the gods. 1458
124 On deification, see also 633 etc. (C 109), 956(E 28)

[134] Cf. descent in 138(D 163).

[135] The verb 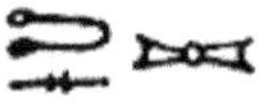 is properly "bind"; but context suggests that det. should have been .

E. Activities of Horus—

RULE

125 Horus has caused that the gods *ḫmᶜ*[136] to Osiris the king, wherever latter goes. 24, 766

126 Horus assigns to (*yp n*) Osiris the king the hearts of the gods. 590

127 Horus has assigned to Osiris the king the gods, so that they have not escaped from (*by*ꜣ *r*) king, wherever latter has taken possession. 615

128 Horus has assigned to Osiris the king all the gods united (*sm*ꜣ·*w*). 1831

129 Horus has brought (*yny*) to Osiris the king all the gods at once, without one of them escaping from (*by*ꜣ *mᶜ*) him. 590, 647

130 Horus (or Geb ?) has brought to Osiris the king the hearts of the gods. 634

131 Horus has caused that the gods follow Osiris the king. 611

132 Horus has given to Osiris the king all the gods, that they may follow king and that latter may control them (*šḫm·k ym·šn*). 620

133 Horus has caused the gods to ascend to Osiris the king, has given them to king to illumine his face. 613, 641, 1659

134 Horus has put Osiris the king in heart (*ḥ*ꜣ*t(y)*) of the gods and has caused that he take every crown (*ṯwy·t*). 613, 648

135 Horus has grasped for Osiris the king the gods, so that they have not escaped from (*by*ꜣ *r*) king, wherever latter has gone. 615

13 6 Horus has caused king to enfold all the gods in his embrace (*ẖnw-ᶜ·wy*) 1632

137 Horus has caused that his children count for (*yp n*)[137] Osiris the king, wherever latter takes possession. 24, 766

138 [Horus has given to Osiris the king his four child]ren to control. 1828

139 Horus has caused that king become a spirit before the spirits, that he gain control before the living (*y*ꜣ*ḫ·k ḫnt(y) y*ꜣ*ḫ·w šḫm·k ḫnt(y) ᶜnḫ·w*). 903

140 Horus, son of Osiris, has put king over his (H's) spirits. 969N

141 Horus of Hieraconpolis has given to king latter's[138] spirits, the jackals. 2011

142 Morning-star, Horus of Dewat etc. is to place king as prince among the spirits, the Imperishable (Stars) who are in the north of the sky. 1220

143 [The son] Horus gives to king latter's *mdw*-staff before the spirits and his *nḥb·t*-scepter before the Imperishable Stars. 1994

144 The two great gods (Horus and Thoth) have supported (*tw*ꜣ) king and establish his seat (*yr·šny yš·t·k*) which is in ———. 1571

145 ⌜Horus⌝ (*Ḥrw*) has put[139] Osiris the king on throne (*nš·t*) of Re-Atum. 1686

146 King is bidden to take the *wrr·t*-crown belonging to the Followers of Osiris, that he may thereby become more a spirit than the spirits, by command of Horus, lord of men (*nb-pᶜ·t*). 1804

147 On rule, see also 634(F 140), 956(E 28), 1294(D 194), 1979(E 60)

[136] Breasted: "bow down."

[137] I.e., ⌜be credited to⌝.

[138] The pronoun here is in 2d per.; but 1294(D 194) has 3d per., referring to Horus.

[139] As Breasted notes, comparing 1692b, *wdn·n·f* here is a scribal error for *wd·n·f*.

E. Activities of Horus—

GLORIFICATION

148 Harakhte causes king to hear his fame (*bꜣ·w*) and his praise (*ḥsw·t*) out of mouth of the two divine enneads. 1449

149 Horus[140] glorifies (*ś·yꜣḫ*) his father (the king). 1712

150 Horus bids king stand, (when) he glorifies him and dispatches him to ascend to sky. 2106

SATISFACTION

151 Horus satisfies (*ś·ḥtp*) king. 897

152 On satisfaction, see also 582 etc. (E 22), 583(F 426), 905(E 90)

MISCELLANEOUS

153 Horus has come as embracer of Osiris the king (*m sḫn·k*). 11

154 Horus has come to embrace Osiris the king (*sḫn·f ṯw*). 575

155 Horus has shone in Osiris the king's presence (*yꜣḫ·n Ḥr ḫr·k*) and in latter's embrace (*m ḥnw-ᶜ·wy·k(y)*). 585

156 Horus has shone again in Osiris the king's presence (*yꜣḫ·n·f ᶜn ḫr·k*). 636

157 Horus has found Osiris the king, and has shone (or 'become a spirit') in (or 'through') him (*yꜣḫ·n·f ym·k*).[141] 612, 648

158 Horus has bound himself to Osiris the king, and has not parted from him (*n wp·n·f yr·k*). 613, 646

159 [Horus unites with (or 'protects'?) Osiris the king (*ḥnm·f ṯw*).][142] 1824

160 When Osiris the king ascends to sky to Re, Horus[143] fraternizes with him. 1016P

161 Horus has caused that the gods fraternize with Osiris the king. 577, 645

162 Horus does for king that which he did for Osiris. 970

163 Horus comes to king to do for him that which he did for his father Osiris, [that he may be more alive than the sky-dwellers and exist] more truly than do those on earth. 2115

164 He who is among the falcons (Horus, named just previously) is to hasten (*śyn*) to king's ka. 1672

165 King's son Horus comes to meet him. 1702

166 Horus bids king stand, Set bids him sit, while his arm is taken by Re (so N; W has: "'take his arm,' says Re"). 473

167 Morning-star, Horus of Dewat etc. is to cause king to sit because of his righteousness (*mꜣᶜ·t*) and to stand because of his reverence (*ymꜣḫ*). 1219

[140] But it is Re who glorifies the king in 795(G 39), and Geb who glorified Osiris as a god in 1013(G 28).

[141] Cf. same phrase with pronouns interchanged in 633 and 1637(C 109). Breasted renders 612: "(that) there is profit for him in thee," and 648: "he has pleasure in thee." Kees (*Opfertanz*, p. 208, n. 23) declares that "shine" is not in early texts written with the *yꜣḫ*-bird which appears in these and the preceding passages.

[142] "Unites with" is favored by parallel phrase "thou art equipped with ⌜him⌝ (*ḫtm·k ym·f*)" in 1827; and *ḫnm*+dir. obj.="unite with" in *Urkunden des äg. Altertums*, IV, 54 (Eighteenth Dynasty). But the similar phrase in Sinuhe R 7 uses the preposition *m*; and *ḫnm*+dir. obj.="protect" several times in Pyr., e.g., in 638 and its parallels.

[143] MN have "Set."

E. Activities of Horus—

168 Given to king is

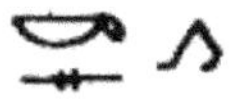

by Horus. 1048

169 ⌜Re, Eastern Horus, and Harakhte, treated as one god,⌝ is to *msms m pḥr·t* (for king). 1088

170 Re is to purify himself for king, and Horus is to adorn (*škr*, for *ḫkr*) himself for him. 1244

171 Horus has not permitted king to *bḫn*. 1633

172 Horus is bidden to betake himself to king's pyramid and to be not distant from it. 1657

173 [(When) king has ascended from lake of life, having purified himself in lake of] the firmament and having become Wepwawet, his son Horus leads him. 1979

174 Horus decks king with the linen (*t3yt(y)·t*) which came forth from him. 2094

175 Horus has set up (*šḫt;* or 'plaited') his arbor (*sḥ*) over king's head (*ḥr ḏ3ḏ3·k;* or possibly 'in king's behalf'); Set has parted (*pšš*) king's ⌜ribs⌝ (*wrm·wt*).[144] 2100

176 Horus provides king with the eye. 19 etc. (F 54), 21 etc. (F 45), 54(F 32), 65(F 55), 104 etc. (F 53), 609(F 64), 742(F 39), 1795(F 40)

177 Other unclassified services of Horus appear in 1684(E 99), 2101–2(D 208), 2213(D 153)

b) Unfavorable

178 Osiris and Horus count not king's heart (*n yp·f yb·k*) and gain no control over his heart (*n šḫm·f m ḥ3t(y)·k*). 145

179 Father Osiris and son Horus have gained no control over king. 146

180 (One) Horus warns away from king's pyramid him whom (another) Horus guards (*twr*)[145] and Set protects (*mk*), Osiris guards and *Ḫrty* protects, Isis guards and Nephthys protects, *Mḫnty-⌜yr·ty⌝* guards and Thoth protects, and the *Ḫ3ty·w* guard and the *Ymy·w-y3w* protect. 1264–65

181 If Osiris, Horus, Set, *Ḫnty-⌜yr·ty⌝*, Thoth, Isis, Nephthys, or the *Ḫ3ty·w* and *Ymy·w-y3w* come with their evil coming, king's pyramid-temple is told not to open its arms to them but to bid them begone. 1267–74

F. Eye of Horus—

I. EPITHETS OF THE EYE

1 *y3b·t*, "left." 451(F 135, F 82), 1231(F 143)
2 *ˁš3·t-wn·w*, "numerous of beings." 901
3 *w3ḏ·t*, "green." 96(F 166), 107, 108(F 166)

[144] Breasted so renders last phrase without question, and interprets that the "ribs" were spread out as supports.

[145] Sethe (*Verbum*, I, p. 144) gives for *twr* "reinigen, schützen." Its parallelism here with *mk* is suggested by Pyr. 350a, where again *Ḥrty* and Osiris are both hostile to king. Gardiner in *Rec. de Trav.*, XXXII, 16 (note on Sinuhe B 11), gives "stand in awe of" as its root meaning.

F. Eye of Horus—

4 *wr·t*, "large." 451(F 135, F 82)
5 *wr·t-bꜣ·w*, "great of honor." 901
6 *wśr·t*, "powerful." 1234(F 152)
7 *wḏꜣ·t*, "sound." 21, 54(F 198), 55(F 78), 900, 1642(F 150)
8 *bnr·t*, "sweet." 100, 111, 591
9 *nḏm·t-śṯ(y)*, "sweet of odor." 1643
10 *nḏś·(t)*, "small." 61 and 88(F 110)
11 *rwḏ·t*, "flourishing." 113(F 206), 614
12 *ḥḏ·t*, "white." 33(F 424), 48(F 115), 96 and 108(F 166)
13 *śwy·t*, uncertain. 600
14 *kꜣꜣ·t*, uncertain. 2087(F 122)
15 *km·t*, "black." 33(F 424)
16 *ṯs·t*, "bound." 1227(F 97)
17 *dšr·t*, "red."[146] 901

II. Magical or Mystic Names of the Eye

18 *wꜣḥ·t-nṯr*, "offering[147] of the god." 614
19 *wꜣg*, a feast.[148] 2185
20 *wr·t-ḥkꜣ·w*, "sorceress (a crown)." 1795
21 *n·t*, "city." 1595
22 *ḥꜣt(y)·t*, "prime oil." 453
23 *ḥknw*, an ointment. 454
24 *grg·wt*, "settlements." 1595
25 *ṯr·t*, "willow-tree." 453
26 *ṯḥn·t*, ⌜= *ṯḥnw*, "olive oil"⌝.[149] 454

III. Relations of Horus to the Eye

27 *Ḥr Šꜣt(y)*, *Ḥr śsmt(y)*, and Eastern Horus, as one god, is bidden to ascend to (*yꜣḳ r*), and does ascend to (*yʿ n*), his large left eye. 452
28 Horus brought (*yny;* or 'carried off'? N adds: 'and supported (*twꜣ*)') the gods by means of the eye. 51
29 Horus performed purification (*yry ʿbw*) for his eye. 1233
30 Alas for (*yhy n*) Horus because of (*n*) his eye; alas for Set because of his testicles. 594
31 "He" (Horus or better Set?) *ytḥ*'ed the eye. 60 etc. (F 106)
32 Horus has taken his eye (and put it) into Osiris the king's forehead (*yṯ·t·n·f r ḥꜣ·t·k*). 54
33 Horus took (*yṯy*) the eye. 107
34 Horus has taken his eye and given (*rdy*) it to king. 1240

[146] Det. shows eye here is the Red Crown.

[147] *Wꜣḥ·t* has loaf and jar determinatives in 101c.

[148] Text has omitted pronominal suffix after *rn*, so this "name" may possibly apply directly to Horus.

[149] Cf. 54(F 368).

F. Eye of Horus—

35 Horus took his eye, rescued (*nḥm*) it from his enemies, (so that) Set ⸢is not profited⸣[150] thereby. 2071

36 "He" (Horus or better Set?) ʿḥ'ed the eye. 105(F 109)

37 Horus ⸢rejoiced over⸣ (*wꜣg*) his eye. 2185

38 *Ḥr Šꜣt(y)*, *Ḥr šsmt(y)*, and Eastern Horus, as one god, is bidden to put ((*w*)*d*(*y*)) his large left eye in his forehead, that he may *ṯrwrw* thereby. 453

39 Horus put that which was in his forehead (eye here representing prime oil) on crown of head (*wp·t*) of his father Osiris. 742

40 Horus has put his eye in Osiris the king's forehead. 1795

41 Horus ⸢assigned (*wḏ*) the eye to⸣ his father ... Orion (here the king). 261(E 21)

42 Horus circles about (*pẖr*) behind (*m sꜣ*; or 'in search of'?) the eye. 670

43 Horus (dwelling) in Osiris the king is bidden to take (*m*) the eye. 19, 21, 55, 831

44 Horus is bidden to take his eye. 1243(E 109), 1614

45 Horus has completely filled (*mḥ*) Osiris the king with his eye. 21, 114, 614, 18[58]

46 Horus filled his empty eye with his full eye (cosmetic). 1682

47 "He" (Horus?) *msꜣ*'ed the eye. 85

48 Horus spoke (*mdw*) concerning the eye in presence of Geb (*ḥr·š ḫr Gbb*). 1643

49 Horus *nn*'ed ('⸢went striding off⸣' or 'did obeisance'?) after the eye. 1595(F 85)

50 The son Horus has rescued (*nḥm*) his eye from Osiris the king's enemy and given it to king. 578

51 Horus has rescued his eye from Set and given it to Osiris the king. 591

52 Horus rescued his eye. See also 1595(F 85), 2071(F 35)

53 Horus has given (*rdy*) (his eye) to king. 104, 105, 107, 108, 109, 1240(F 34)

54 Horus has given his eye to Osiris the king. 19, 578(F 50), 591(F 51), 614, 634, 641, 1805

55 Horus has given his eye into hand (ʿ) of Osiris the king. 65

56 Horus gave his eye to Osiris. 1643

57 Horus has not permitted (*rdy*) his eye (Egypt or some part thereof) to hearken to Westerners, Easterners, Southerners, Northerners, nor those in the midst of (*ḥry·w-yb*) the land. 1588

58 Horus picked up (*ḥꜣm*) the eye. 93

59 Horus rejoiced (*ḥʿy*) at approach (*ḥšfw*) of his eye (when) it was given (back) to him before his father Geb. 977

60 Horus is praised (*ḥkn*) thereby. 454(F 89)

61 Horus is satisfied with (*ḥtp ḥr*) his eye. ⸢58⸣, 2072

62 Horus was satisfied with his eye. 59

[150] Sethe (quoted by Breasted) thinks

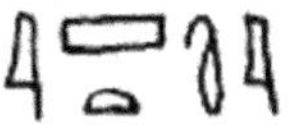

a denominative verb from *yš·t*.

F. Eye of Horus—

63 Horus (dwelling) in Osiris the king is bidden to equip (*ḥtm*) himself with the eye. 21

64 Horus has equipped Osiris the king with his eye, has united (*ś·dmy*) his eye to king. 609

65 "He" (Horus or better Set?) *ḥnf*'ed the eye. 76 etc. (F 113)

66 Horus falls (*ḫr*) because of (*n*) his eye; the bull (so W; T has 'Set') collapses (*sbn*) because of his testicles. 418

67 Horus falls because of (*ḫr*) his eye; Set suffers (*pꜣs*)[151] because of his testicles. 679

68 "He" (Horus?) ⸢*ḫẖr*⸣'ed the eye. 108

69 "He" (Horus) took possession of (*ẖnm;* or 'protected') the eye. 51

70 Horus goes to the eye (*sy r·ś*). 31

71 "He" (Horus or better Set?) *sḫs*'ed the eye. 109(F 114)

72 Horus recognized (*śyꜣ*) his eye in the Prince-house that is in Heliopolis (*ḥt·t-śr ymy·t Ywnw*). 1614

73 "He" (Horus?) handed over the eye (*y·ś·(y)p·t·n·f*). 109

74 "He" (Horus?) ⸢—ed upon⸣ his eye (*y·śfkk·t·n·f ḥr·ś*). 51

75 "He" (Horus?) *ś·swn*'ed the water ⸢by means of⸣ (*m;* or 'from' or 'in' or 'into'?) the eye. 88

76 "He" (Horus or better Set?) *ś·šd*'ed the eye. 108(F 116)

77 Horus (dwelling) in Osiris the king is bidden to unite (*ś·dm(y)*) the sound eye to his face. 55

78 Horus united his sound eye. 55

79 Horus has united his eye to Osiris the king. 609(F 64)

80 Horus goes to the eye (*šm yr·ś*). 83

81 Horus went about (*šn*) in the protection of his eye. 195 etc. (C 92)

82 *Ḥr Šꜣt(y)*, *Ḥr śsmt(y)*, and Eastern Horus, as one god, is asked to receive (*šsp*) his large left eye from king in sound condition, with its waters (*mw*), its colors (*ṯr·w*), and its ⸢throats⸣ (⸢*ḥt(y)·w(t)*⸣) in it. 451

83 Horus took (*šd*) the eye from Set. 95

84 Horus built (*ḳd*) the eye. 1589(F 92)

85 Horus settled (*grg*) the eye (Egypt or some part thereof), ⸢went striding off⸣ (*nn(yw)·f;* Breasted takes as *nyny*, 'did obeisance') after it, and rescued (*nḥm*) it from every green thing which Set did against it. 1595

86 Horus settled the eye. See also 1589(F 92)

87 Horus supported (*twꜣ*) the gods by means of the eye. 51(F 28)

88 Horus is to *ṯrwrw* thereby. 453(F 38)

89 *Ḥr Šꜣt(y)*, *Ḥr śsmt(y)*, and Eastern Horus, as one god, sparkles (*ṯḥnḥn*) thereby among the gods, and is praised (*ḥkn*) thereby. 454

90 Horus besought his eye from (*dbḥ mꜥ*) Set. 65

91 Horus adorned (*ḏbꜣ*) his eye (Egypt or some part thereof) with both his arms (*ꜥ·wy*). 1588, 1596(C 94)

92 Horus adorned the eye (as above), built (*ḳd*) it, settled (*grg*) it. 1589

93 Others provide Horus with his eye. See 451(F 135), 535(D 198), 946(F 98), 1235 etc. (F 203), 1239(F 197), 1643(F 173)

151 So Sethe, *Verbum*, I, p. 158.

F. Eye of Horus—

94 Other connections of Horus with his eye. 19 etc. (F 309), 61 etc. (F 227), 72(F 243), ⸢74⸣(F 181), ⸢86⸣(F 182), 737(F 149), 901(F 263), 1407(F 259), 1589–92(F 279), 1593–94(F 291), 1756(F 316), 2072(F 286, F 308, F 289), 2088(F 318), 2090(F 196), 2091(F 120)

IV. RELATIONS OF OTHERS TO THE EYE

95 The spirits (*y᾽ḫ·w*). 57(F 235), 579(F 159)

96 Osiris. Besides references to king (F 130 ff.), see 742(F 39), 1643(F 56)

97 *M᾽-ḥ᾽·f*, the ferryman, called also *Ḥr·f-ḥ᾽·f*, is bidden to bring (*yny*) for king the bound (*ṯs·t*) eye of Horus which is in Field of ⸢Rowers⸣ (*śḫ·t-ḫnn·w*),[152] namely (the barque) "Eye of Khnum." 1227

98 The ferryman (*mḫnty*) is bidden to bring to Horus his eye and to Set his testicles. 946

99 The gods (*nṯr·w*) ⸢crushed⸣ (*p᾽*) eye of Horus, that they might eat (*wšb*) there(of). 117

100 The gods who ferry over on wing of Thoth to yonder side of Lily Lake to east side of sky to speak with (*md·t ḫft*) Set about (*ḥr*) eye of Horus are entreated to ferry king over with them to same place for same purpose. 595–96

101 The gods fear before (*nr n*) eye of Horus. 1755

102 The gods. See also 51(F 28), 634(F 140), 737(F 149), 1147(F 236), 1231(F 143), 2075(F 160)

103 Re. 2090(F 319)

104 ⸢The hoers⸣, ⸢"they who wear the skin"⸣ (former written

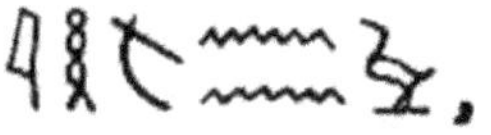

to which

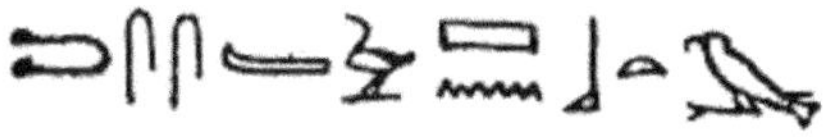

is parallel), have swallowed (*ʿm*) the eye. 118

105 Khnum. 1227(F 97)

106 "He" (Set?) ⸢confined⸣ (*ytḥ* for *rtḥ*) the eye. 60, 73, 77, 78, 86

107 Set took (*yṯy*) the eye. 1233, [2213]

108 "He" (Set?) is not to swallow (*ʿm*) the eye. 92(F 179)

109 "He" (Set?) ⸢withheld⸣ (*ʿḥ*) the eye.[153] 105

110 Set ate of (*wnm m*) the small eye of Horus. 61, 88

111 The eye was wrested (*hp*) from Set. 36, 39

112 Set swallowed (*ḫnp*) the eye. 1839(F 218)

113 "He" (Set?) ⸢swallowed⸣ (*ḫ᾽nf* and *ḫnf* resp.) the eye. 76, 95

114 "He" (Set?) ⸢carried it off⸣ (*sḫs*).[154] 109

115 [Finger of Set causes] the white eye of Horus [to see (*ś·m᾽᾽*)]. 48

[152] Or perhaps "Field of Conflict"; for, while det. of *ḫnnw* is a ship in P, in M it is a (mutilated) man with weapon. *Cf.* "that place where they fought" in 1242(F 204).

[153] *Cf.* 99(F 180).

[154] *Cf.* 97(F 183).

F. Eye of Horus—

116 "He" (Set?) ⸢took⸣ (*š·šd*)[155] the white and the green eye of Horus. 108
117 Set ⸢tramples⸣ (*ty*)[156] the eye. 73
118 Set. See also 20(F 288), 48(F 276), 65(F 90), 84(F 246), 95(F 83), 591 (F 51), 594(F 278), 595–96(F 100), 1242(F 204), 1407(F 259), 1593–94(F 291), 1595(F 85), 1742(F 226), 2071(F 35)
119 Shu is bidden to bear (*wṯs*) eye of Horus to sky, to the *šḥdw* of the sky. 2091
120 Shu is bidden to go (*sy*) as[157] one who shall row (*šš3w·t(y)·f(y)*) Horus ⸢on account of⸣ (*ḥr*) his eye. 2091
121 Geb gives (*rdy*) to king that which (was) in forehead of Horus. 139
122 Geb has lifted up (*š·ṯs*) the *k33·t* eye. 2087
123 Geb. See also 977(F 59), 1643(F 48)
124 Thoth is to bring (*yny*) king bearing (*ḥr*) the eye. 58
125 Thoth is bidden to set on (*(w)d(y)*) for king the eye of Horus (king's head).[158] 830
126 Thoth is bidden to ascend (*pr*) to king bearing the eye. 58
127 Thoth avenged (*nḏ*) the eye. 2213
128 Thoth is bidden to give (*rdy*) the eye to king. 58
129 Thoth. See also 43(F 294), 594(F 278), 976(F 258)
130 King is the eye of Horus. 698(F 280), 976(F 300), 1147, 1460(F 409)
131 Osiris the king is bidden to unite (*yʿb*) the eye to his mouth. 60, 72
132 King is not left boatless (*ywy*), (for) he possesses (*ḥr*) the eye. 1429
133 King *ywg*'s after the eye. 1067(F 229)
134 King is bidden to assign (*yp*) the eye to himself. 100, 111, 591
135 King brings (*yny*) to *Ḥr Š3t(y)*, *Ḥr šsmt(y)*, and Eastern Horus, considered as one god, latter's large left eye. 451
136 King brings to Horus his eye. 535(D 228)
137 Osiris the king is to betake himself to (*ys3 r*) the eye. 82
138 Osiris the king is to take (*yṯy*) the eye. 67, 1838
139 Osiris the king takes the eye (wine) to his mouth. 36
140 Osiris the king is to take *wrr·t*-crown before the gods by means of the eye. 634
141 King is to take *wrr·t*-crown by means of the eye. See also 737(F 149), 2075(F 160)
142 King has not swallowed (*ʿm*) eye of Horus nor a member (*ʿ·t*) of Osiris, that he should die because of either. 1450

[155] This verb in both 96(F 184) and 108(F 116) may have nothing to do with *šd*, "take," for writing and subject alike differ, e.g., in 95(F 83).

[156] Cf. the reduplicated form *tyty* in Erman's *Glossar*. Comparison of 73(F 117) and 60(F 185) shows that Set is the offender in both cases.

[157] The writing

in 2091(F 120) seems to be a scribal error, for sense requires action favorable to Horus. In 2090(F 196) *m* does in fact stand alone.

[158] Cf. 639 and 10b.

F. Eye of Horus—

143 King stands at (ʿḥʿ r) that left eye of Horus, where'in' (m; or 'by' or 'concerning which')[159] the word of the gods is heard. 1231

144 King has purified himself (wʿb) with the eye of Horus; his injury (śḏb) has been removed by the two falconesses of Osiris (ḏr·ty-Yśyr, Isis and Nephthys). 308, 312

145 Osiris the king opens (wp) his mouth with the eye of Horus (wine). 36

146 King is to open his mouth with the eye. 39, 63, 92, 93, 106

147 Osiris the king is to open his mouth with that which lays hold of him (mḥ·t ym·k, the eye, here wine). 36

148 King becomes (wn) Presider over the Westerners by means of the eye. 139(F 161)

149 King is bidden to clothe himself in (wnḫ m) the eye of Horus which is in T³y·t, that it may be king's ky·t in presence of the gods and the means of his recognition (m·śy³·t) by them, that he may take (yṯy) wrr·t-crown by means of it from the gods and from Horus, lord of men (nb-pʿ·t). 737

150 King is bidden to put on (wnḫ) the sound eye of Horus which is in T³y·t. 1642

151 King is bidden to put on eye of Horus, to receive (šsp) it upon himself, that it may unite (dmy) with his flesh and that he may ascend (pr) thereby and the gods see him decked (ḏb³) therewith. 844-45

152 King puts on (wṯs) the White Crown (ḥḏ·t) the powerful (wśr·t) eye of Horus there. 1234

153 Osiris the king is bidden to put ((w)d(y)) the eye within himself (mm·k), that "he" (king's enemy) may fear before him (nr·f n·k). 249

154 Osiris the king is bidden to put the eye within himself, that his every enemy may fear before him. 614

155 Osiris the king is to put the eye unto himself (ḫr·k). 1798, [18]08

156 King is to put the eye ('butter or cheese') in his mouth. 31

157 Osiris the king has put the eye in his mouth. 77

158 King puts on crown of his head (wp·t) that which was in forehead of Horus (eye here representing prime oil). 742T

159 Osiris the king is to become a soul (b³) and gain control (śḥm) before the spirits by means of the eye. 579

160 King becomes a soul, gains control, receives obeisance (w³š), and takes (yṯy) wrr·t-crown among the gods, by means of the eye. 2075

161 King becomes a soul, gains control, and becomes (wn) Presider over the Westerners, by means of that which (was) in forehead of Horus. 139

162 Osiris the king 'crushes' (p³) eye of Horus (cakes) in order to eat. 35, 78

163 King is to ascend (pr) thereby. 845(F 151)

164 Osiris the king is to carry (f³) the eye to his face. 61

165 King is bidden to take (m) the eye of Horus. 19, 31, 35, 36, 38, 39, 40, [43], [48], 51, 59, 60, 61, 64, 72-74, 76-78, 80, 82-89, 92-95, 97-100, 105-10, 117, 1756, 1807, 1839, 1840, 2033

166 Osiris the king is bidden to take both the white and the green eye of Horus. 96, 108

---

[159] Cf. 595-96(F 100).

F. Eye of Horus—

167 Osiris the king is to see (*m3·k*) by means of the eye of Horus.[160] 641, 1805, 1807, 1808
168 King speaks with (*mdw·f ḫft*) Set about eye of Horus. 596(F 100)
169 The eye was rescued (*nḥm*) for king. 39, 40
170 King is to *nḫḫ* the eye unto himself. 21, 115, 1068, 1881
171 King is to *nḫḫ* (the eye) unto himself. 100, 112, 216, 591
172 Osiris the king ⸢fed upon⸣ (*nšbšb ḥr*) the eye. 64
173 ⸢King⸣ has given (*rdy*) the eye (back) to Horus to equip (*ḥtm*) his face therewith. 1643
174 The eye is presented (*ḥnk*) to king. 1881
175 King sought (*ḥḥ*) the eye in Pe, found (*gm*) it in Heliopolis. 1242
176 King is to be satisfied with (*ḥtp ḥr*) the eye. ⸢58⸣, 59
177 King is to equip (*ḥtm*) himself with the eye. 40, 901
178 King is to equip his face with the eye (incense). 19, 20, 29
179 Osiris the king is to prevent (*ḫw*) that "he" (Set?) swallow (*ʿm*) the eye. 92
180 Osiris the king is to prevent that "he" (Set?) ⸢withhold⸣ (*ʿḥ*) the eye. 99
181 Osiris the king is to prevent that "he" (Horus?) suffer (*p3s*)[161] because of (*ḥr*) the eye. 74
182 Osiris the king is to prevent that "he" (Horus?) suffer from the eye (*mn·f ś(y)*). 86
183 Osiris the king is to prevent that "he" (Set?) ⸢carry off⸣ (*sḫs*) the eye. 97
184 Osiris the king is to prevent that "he" (Set?) ⸢take⸣ (*ś·šd*)[155] the white and the green eye of Horus. 96
185 Osiris the king is to prevent that "he" (Set) ⸢trample⸣ (*ty*) the eye.[156] 60
186 King advances (*ḫnt*) possessing (*ḥr*) the eye of Horus, and sits upon seat of the gods. 1241
187 Osiris the king embraces (*sḫn*) the eye. 80
188 King is in embrace of eye of Horus. 600, 1242
189 King punished (*ss*) his enemies by means of the eye. 1240(F 429)
190 Osiris the king is to assign (*y·ś·(y)p*) the eye to himself. 87
191 [Osiris the king is not to let] go of ([*ś·f*]*ḫḫ*) the eye. 43
192 Osiris the king is not to *ś·ḥbnbn* the eye. 76
193 Osiris the king is to drive it back (*ś·ḥm*) from himself. 59
194 King is to gain control (*śḫm*) by means of the eye. 139(F 161), 579(F159), 2075(F 160)
195 King is to ⸢assume⸣ (*ś·ḫt*) the eye to himself. 46, 100, 111, 591
196 King is to go (*sy*) as[157] one who shall row (*m śš3w·t(y)·f(y)*) Horus ⸢on account of⸣ (*ḥr*) his eye. 2090
197 King fastens on (*ś·ts*) for Horus that which came forth from latter's head. 1239
198 King is to unite (*ś·dm(y)*) to his face the sound eye of Horus (cosmetic). 54

[160] In 610 and 1806(E 43), after Horus has bestowed his eye on the king, he opens "thy (the king's, title having been transferred by the previous act) eye, that thou mayest see thereby."

[161] Used of Set in 679(F 67).

F. Eye of Horus—

199 King is to follow (*šmś*) eye of Horus to sky, to the *šḥdw* of the sky. 2090
200 King is to receive (*šsp*) his bread, even the eye of Horus. 63
201 King is to receive the eye upon himself. 844(F 151)
202 King is to take (*šd*) the eye. 1354
203 King ascending to sky takes eye of Horus to latter. 1235, 1237, 1239
204 King took the eye of Horus from head of Set in that place where they fought. 1242
205 King's heart is to be refreshed by possession of the eye (*ḳb ḥr·ś*, libation). 22
206 Osiris the king is to ⌜*kꜣp*⌝[162] the flourishing eye within himself (*ym·k ym·k*), that his enemy may fear before him (*nr n·k*). 113
207 King found (*gm*) the eye. 1242(F 175), 2089, 2090
208 Osiris the king is to ⌜satisfy⌝ (*dꜣp*) himself with the eye. 110
209 King is the ⌜—⌝ falcon (*byk ngg*) encircling (*dbn*) the eye of Horus of Dewat (*Ḥr ḥry-yb D(w)ꜣ·(t)*). 1959
210 King tastes (*dp*) the eye (a loaf). 38
211 The gods are to see king decked (*ḏbꜣ*) with the eye. 845(F 151)
212 Others than Horus[163] provide king with the eye. 18 etc. (F 220), 20 etc. (F 215), 22(F 216), 58(F 124, 126, 128), 139(F 121), 216(F 214), 742MN (F 217), 830(F 125), 1068(F 219), 1227(F 97), 1755(F 221)
213 Other connections of king with the eye. 12(F 305), 18(F 312), 19(F 314), 20(F 274, 288), 20 etc. (F 310), 38(F 273), 40(F 241), [43] etc. (F 296), 57(F 269, 235), 64 etc. (F 306), 67(F 262), 72(F 295), 79 (F 304), 87 (F 281), 93(F 302), 94(F 275), 104(F 299), 116(F 283), 758(F 228, 339), 846(F 270), 901(F 263), 976(F 282), 1067(F 229), 1241(F 311), 1407 (F 259), 1754(F 313), 1800–1801(F 298), 2033(F 256), 2076(F 285)
214 ⌜Priest⌝ (1st per. throughout the following group) has assigned (*yp*) the eye to king. 216
215 ⌜Priest⌝ brings (*yny*) the eye to king. 20, 22, 31, 54, 846, 1794, 2074
216 ⌜Priest⌝ brings the eye (libation) under king's soles (*kb·wy*). 22
217 ⌜Priest⌝ puts (*(w)d(y)*) on crown of king's head (*wp·t*) that which was in forehead of Horus (eye here representing prime oil). 742MN
218 ⌜Priest⌝ rescued (*nḥm*) the eye from Set (after) latter had swallowed (*ḫnp*) it. 1839
219 ⌜Priest⌝ gives (*rdy*) the eye to king. 1068
220 ⌜Priest⌝ has given the eye to Osiris the king. 18, 29, 66, 216, 1808
221 ⌜Priest⌝ has decked (*ḏbꜣ*) Osiris the king with the eye (linen). 1755
222 "They" (obscure) come ⌜—⌝ by means of (or 'in' or 'from'?) the eye of Horus (*yw·śn*

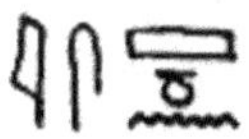

ym·ś). 89

[162] Probably scribal error,

for *d*; cf. 249 and 614(F 153–54).

[163] See latter summarized under E 176.

F. Eye of Horus—

223 "They" spewed out (*bš*) the eye. 92
224 "They" ⸢eat⸣ (*y·nśb·t·śn*) the eye. 104
225 "They" ⸢ate⸣ (*y·nśb·t·n·śn*) the eye. 98
226 "Put ye" (read

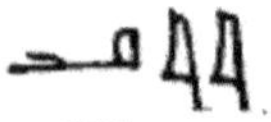

as *rdy·y*, impv. pl.) the eye of Horus upon wing of his brother Set. 1742
227 "They" *ḥḥm*'ed eye of Horus against him. 61, 89

V. Actions, Circumstances, and Qualities of the Eye

228 The eye comes (*yy*) to king and addresses (*mdw*) him. 758
229 The eye comes (*yw*) to king at first of the ten(-day period), while he *ywg*'s after it. 1067
230 The eye is conceived (*ywr·t*). 698(F 280)
231 The eye is ⸢greater than⸣ (*ʿꜣ r*) king. 116(F 283)
232 The eye lives (*ʿnḫ*). 2050(F 240)
233 The eye is distant from king (*wꜣꜣ·ś ḥr·k*). 1881
234 The eye (king's pyramid and mortuary temple) is pure (*wʿb*). 1277
235 The eye opens (*wp*) king's way before (*ḫnt*) the spirits. 57
236 The eye is mighty against men (*wśr·t yr r(m)ṯ·(t)*), victorious (*nḫt*) against gods.[164] 1147
237 The eye spits not (nor) spews (*n wg·ś bšš·ś*). 1460P(l. 661)
238 The eye spat and spews (*wgy·t bšš·t*). 1460P(ll. 774–75)M
239 The eye is sound (*wḏꜣ*). 21, 451(F 82)
240 The eye of Horus which is in Heliopolis is sound and lives. 2050
241 The eye has not escaped from (*byꜣ r*) king. 40
242 The eye spews (*bšš*) or spews not. 1460(F 237–38)
243 The eye ⸢purged⸣ (*bd*) mouth of Horus. 72
244 In (*m*) Heliopolis. 1242(F 175), 2050(F 240)
245 In Pe. 1242(F 175)
246 In forehead (*ḥꜣ·t*) of Horus and that of Set respectively. 83–84
247 In Houses of Red Crown (*ḥt·wt-n·t*). 56(F 264)
248 In Field of ⸢Rowers⸣ (*śḫ·t-ḫnn·w*). 1227(F 97)
249 In (city of) *Tꜣy·t*. 737(F 149), 1642(F 150), 1794, 2074[165]
250 In *Dp*. 56(F 264)
251 The eye is to see (*mꜣ*) Horus. 2088(F 318)
252 The eye is that which lays hold of king (*mḥ·t ym·k*). 36(F 147)
253 Born (*mś·t*) every day. 698(F 280)
254 Is king's means of recognition (*m·śyꜣ·t*). 737(F 149)
255 Addresses (*mdw*) king. 758(F 228)
256 Belongs to Osiris the king (*n(y)-ṯw ś(y)*, in literal instead of the usual inverted sense). 2033
257 The eye *nbꜣbꜣ*'s. 98, 104

[164] Or "more ... than ..." in both cases, as Breasted renders.

[165] Emend according to 1794.

F. Eye of Horus—

258 The eye *nbḏbḏ*'s upon wing of Thoth in east (left) side of ladder of the god. 976

259 The eye has *nḥḥ*'ed unto ⌜king and Horus⌝; has not been given to the attacker (*ḏnd*), Set. 1407

260 ⌜Victorious against⌝ (*nḫt r*) gods. 1147(F 236)

261 The eye trickles (*ndfdf*) on the *ḏnw*-bush. 133, 695

262 Avenges (*nḏ*) Osiris the king. 67

263 The eye is to avenge king as it avenges Horus. 901

264 Eye of Horus which (is) in *Dp*, eye of Horus which (is) in Houses of Red Crown (*ḥt·wt-n·t*),[166] wakes (*rś*) in peace, having received the *yr·wt* that adorned the *wr*-ᶜ-chapel. 56

265 Causes that the Two Lands bow (*rdy·ṯ kśy*) to king as they bowed to Horus, that they fear before (*nr n*) king as they feared before Set. 57

266 Serves to inspire fear in king's enemy. See 113(F 206), 249(F 153), 614(F 154)

267 Fear (*śnḏ*) arose because of (*ḫpr ḥr*) the eye. 1040

268 The eye *ḫbnbn*'s. 94, 107

269 Sits before (*ḥmś·ṯ ḫft*) king as his god. 57

270 Not distant from (*ḥr r*) king for ever and ever. 846

271 Protects (*ḫw*) king from . . . Set. 20(F 288)

272 Falls (*ḫr*) in east side of sky. 947(F 277)

273 Is king's *sḫnty*. 38

274 The eye (incense) is to purify (*ś·(w)ᶜb*) king. 20

275 The eye *śnw*'s not against king. 94

276 [Illumines (*ś·ḥḏ*) tip of fin]ger of Set. 48

277 The eye, and king with it, turns (*śṯp*) and falls (*ḫr*) in east side of sky. 947

278 When it has seen Thoth, the eye turns and falls upon Thoth's wing in yonder side of Lily Lake, to protect itself (*y·nḏ·ś ḏ·t·ś*) from Set. 594

279 Eye of Horus (Egypt or some part thereof) hearkens to (*śḏm n*) Horus only, doing for him everything which he says to it wherever he goes, carrying (*fꜣ*) to him the swamp-waters, all the wood, the food, the drink-offerings, everything, which is or shall be in it.[167] 1589–92

280 King is this eye of Horus[168] which sleeps (*śḏr·t*), is conceived (*ywr·t*), and is born (*mś·t*) every day. 698N

281 The eye cuts not (*šᶜ*) against Osiris the king. 87

282 King goes (*šm*), going as the eye of Horus (goes). 976

283 Eye of Horus, as incense, is higher and greater than (*ḳꜣ·ṯ ᶜꜣ·ṯ r*)[169] king. 116

284 The eye is king's *ky·t*. 737(F 149)

285 Is upon king (*tpy·t·k*). 2076(D 144)

286 Unites with (*dmy m*) Horus. 2072

287 Is to unite with (*dmy r*) king, with his flesh. 844(F 151)

[166] These "eyes" stand parallel to *Tꜣy·t* and *Tꜣyt(y)·t*.

[167] Parallel passage 1599–1602 has Nut (mentioned in 1596) hearkening to and serving the king.

[168] TP have "Re."

[169] Or "is high and great upon." Breasted suggests both.

F. Eye of Horus—

288 Eye of Horus (incense) removes (*dr*) the (odor of) king's moisture (*rḏw*) and protects him from the flood of the ˹region˺ of Set (*ḥw·š ṯw mʿ ȝgb n ʿ n Stš*). 20

289 Attack (*ḏnd*) of eye of Horus falls (*ḫr*) against his enemies. 2072

290 See also 2087(D 193) and the epithets in F 1–17

VI. PARTS AND ACCESSORIES OF THE EYE

291 The doors (*ʿȝ·w*) which are upon the eye of Horus (Egypt or some part thereof) stand like *Ywn-mw·t·f*, and open not (*wn*) to Westerners, Easterners, Northerners, Southerners, nor those in the midst of the land, but (only) to Horus; for Horus made and erected them and rescued (*nḥm*) them from every ill which Set did against them. 1593–94

292 Beings (*wn·w*). 901(F 2)

293 Falcon (*byk*) came forth from (or 'as'?) eye of Horus. 1843

294 [Thoth saw the waters (*mw*)] which (were) in the eye of Horus. 43

295 Osiris the king is to unite (*yʿb*) to himself the waters which (are) in the eye. 72

296 Osiris the king is to take (*m*) the waters which (are) in the eye. [43], [47]

297 Waters. See also 451(F 82)

298 King is to fill himself with the ointment (*mḏ·t*) which came forth from the eye, that it may fasten (*ṯs*) his bones, unite (*dmḏ*) his members, bring together (*šȝk*) his flesh, and loose (*šfḫ*) his evil sweat to earth. 1800–1801

299 King is to take (*m*) ˹adornment˺ (˹*nḫb·t*˺) of the eye. 104

300 Foot (*rd*) of the eye of Horus (eye here is the king) is limited (*ḏr*) wherever it is.[170] 976

301 Forehead (*ḥȝ·t*). 2090(F 319)

302 Osiris the king is to take (*m*) pupil (*ḥwn·t*)[171] which is in the eye of Horus. 93

303 ˹Throats˺ (˹*ḥt(y)·w(t)*˺). 451(F 82)

304 Osiris the king is to take *ḫpḫ* of the eye of Horus. 79

305 Osiris the king's mouth is opened (*wp*) by *ḫpḫ* of the eye. 12

306 Osiris the king is to take the *šwt·t* of the eye of Horus. 64, 81

307 A band of green and of dark red linen (*ššd pw ny wȝḏ·t n(y) ydmy*) was spun (*šṯȝ*) from the eye. 1202

308 The odor (*šṯ(y)*) of his eye is upon (*r*) Horus. 2072

309 Horus *pḏ*'ed with (*m*; or 'in'?) odor of his eye. 19, 1754

310 Odor of the eye of Horus is upon (*r*) king. 20, 1803, 2074, 2075

311 Odor of the eye of Horus is upon king's flesh. 1241

312 Odor of the eye of Horus (incense) *pḏpḏ*'s[172] to king. 18

313 Osiris the king is to take upon himself the odor of the eye. 1754

314 Osiris the king is to equip (*ḫtm*) himself with odor of the eye (incense). 19

315 Odor. See also 1643(F 9)

[170] Similar passages in 622 and 625 refer to king directly, and declare that his foot is *not* limited.

[171] Cf. 206(A 57).

[172] Breasted: "˹adheres˺."

F. Eye of Horus—

316 Its ⌜*tnf*⌝ was made for the eye of Horus.[173] 1756
317 Colors (*ṯr·w*). 451(F 82)
318 Head (⌜*d*ꜣ*d*ꜣ⌝) of eye of Horus was given, that it might see (*m*ꜣ) Horus. 2088
319 A head was given to the eye of Horus, and a forehead (*ḥ*ꜣ*·t*) was made for it out of crown of Re's head (*wp·t*). 2090

VII. SYMBOLISM OF THE EYE OF HORUS. IT REPRESENTS

320 ꜣ*mś*-scepter.[122] 43
321 (Wine of) Nebesheh (*Ym·t*).[174] 93WN
322 Wine (*yrp*). 36, 39
323 Wine of Nebesheh (*yrp-Ym·t*). 93T
324 Wine of the Delta (*yrp-*⌜*ydḥ·w*⌝). 92, 106
325 Wine of ⌜—⌝ (*yrp-*ʿ*bš*). 92T
326 Wine of ⌜—⌝ (*yrp n ḫ*ꜣ*mw*). 93T
327 Wine of Pelusium (*yrp-Śyn*).[175] 94T
328 Every sweet ⌜thing⌝ (⌜*yḫ·t*⌝ *nb·(t) bnr·t*). 100, 111
329 *Yšd*-fruit. 95
330 Barley (*yt*). 97
331 ⌜Incense⌝ (*yd·t*). 77
332 (Wine of) ʿ*bš*. 92WN
333 Seed-grain (ʿ*g·t*).[176] 109
334 Green cosmetic (*w*ꜣ*dw*). 54
335 *W*ʿ*ḥ*-grain.[177] 99, 105
336 Rolls of linen (*wnḫ·w*). 57
337 *Wr·t*-⌜cakes⌝. 103
338 *Wrr·t*-crown. 845
339 King's soul (*b*ꜣ) and control (*sḫm*). 758
340 *B*ꜣ*b*ꜣ*·t*-⌜grain⌝. 98, 104
341 The olive-tree that is in Heliopolis (*b*ꜣ*ḳ·t ymy·t Ywnw*). 118
342 *Bd*-natron.[178] 72

[173] Text reads

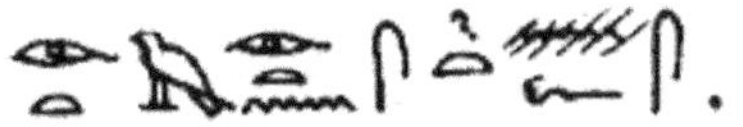

Breasted renders: "... which ⌜Set⌝ made ..."

[174] On *Ym·t* as Nebesheh see Petrie, *Nebesheh and Defenneh*, p. 6 and Pls. X, XI.

[175] On *Śyn* as Pelusium see Spiegelberg in *Zeitschrift für äg. Sprache*, XLIX, 81. These and many following items of the offering-ritual are discussed by Bollacher in von Bissing, *Mastaba des Gem-ni-kai*, II, 37–40, as well as in Budge's books based on Maspero and Dümichen (cf. notes *passim*).

[176] See references in *Gemnikai*, II, 39. Bollacher there reads ʿꜣ*g·t*; but Erman (*Gram.*[3]) gives under Schrifttafel S 47 the value ʿ, not ʿꜣ.

[177] *W*ʿ*ḥ*-grain is made into ʿꜣ*b·t*; see Newberry, *Rekhmara*, Pl. XII and p. 35. Other references also are in *Gemnikai*, II, 39.

[178] Budge (*Liturgy of Funerary Offerings*, 111) calls *bd* "incense," though the vignette which he reproduces shows

as determinative.

F. Eye of Horus—

343 *P3·t*-cakes. 78
344 *P3·t-wd3·t*-cake. 35N
345 An offering-cake (*p3·t n·t wdn*). 35W
346 A mortuary offering (*pr·(t r) ḫrw*). 58
347 A royal mortuary offering. 59
348 *Psn*-loaves. 74
349 ⸢Liver⸣ (*m(y)s·t*). 82, 88
350 Water (*mw*). 10
351 *Mnw*-jars. 33
352 Dove (*mnw·t*). 86
353 Raiment of the god (⸢*mnḫ·t*⸣-*nṯr*). 42
354 King's pyramid (*mr*) and mortuary temple (*ḥt·t-nṯr*). 1277(F 234)
355 Black cosmetic (*m·šdm·t*). 54
356 *Mdw*-staff. ⸢43⸣
357 *Nbś*-fruit. 98, 104
358 *Np3·t*-fruit or grain. 87, 109
359 Spleen (*nnšm*). 83
360 *Nḥb·t*-scepter. ⸢43⸣
361 *Nẖnm*-oil. 51
362 Natron (*nṯry*). 23
363 Every fruit (*rnp·t nb·(t)*). 100, 111
364 *Rnn-wt·t*-linen. 1755, 1794
365 ⸢—⸣ (*h3ṯr* or *ṯr*). 66
366 A breast of meat (*ḥ3·(t)-y(w)f*). 84
367 (Wine of) *ḥ3mw*. 93WN
368 Prime olive oil (*ḥ3ty·t-ṯḥnw*).[179] 54
369 A limb (*ḥ*ʿ). 83
370 *Ḥbnn·t*-⸢fruit⸣. 76, 94, 107
371 A (drink)-offering or libation (*ḥnk·t*). 100, 112
372 Beer (*ḥk·t*). 39, 40, 61
373 The offering of the broad hall (*ḥtp-wśḫ·t*). 59, 103
374 A royal offering (*ḥtp-(ny)-św·t*). 58, 59
375 Temple-income (*ḥtp-nṯr*, with food ⸢determinatives⸣). 115
376 The White Crown (*ḥḏ·t*). 1234(F 152)
377 An altar (*ḫ3(w)·t*). 58
378 A joint of meat (*ḫpš*). 12, 79
379 *Ḫnfw*-⸢fruit⸣. 76, 95
380 *Ḫnmś*-beer.[180] 61, 89
381 ⸢*Ḫrḫnf*⸣-⸢fruit⸣.[181] 108

[179] On *ṯḥnw* as "olive," see Newberry in *Ancient Egypt*, 1915, pp. 97–98. This is one of seven oils prominent in the Old Kingdom. The Metropolitan Museum (New York) possesses in its Old Kingdom collection a rectangular alabaster tablet with the seven oil-cups each properly inscribed. There are three similar tablets in the British Museum; see Budge, *Book of Opening the Mouth*, I, xiii–xiv.

[180] Pyr. writing would indicate *ḫnśm*; but the true reading is proved by variants listed by Dümichen in *Grabpalast des Patuamenap*, I. Abt., Pl. XXIV, No. 92. Cf. also the word *3mś*, with *ś* written before *m* in Pyr. 274 and 1166.

[181] Probably intended for *ḫnfw* (F 379); note relation of *ḥbnn·t*-series (F 370).

F. Eye of Horus—

382 A *s·t*-goose. 85
383 A libation (*sꜣṯ*). 1840
384 Wheat (*swt*). 97
385 ⌜Butter or cheese⌝ (*srw*).[182] 31
386 A joint of meat (*sḫnw*).[183] 38, 80
387 A *šꜣ·t*-garment. 41
388 A joint of meat (*šw·t*). 64, 81
389 *Šfṯ*-oil. 51
390 (Wine of) Pelusium (*Šyn*).[175] 94WN
391 Incense (*šnṯr*). 18, 19 (cf. 1644a), 20–21a (cf. 1644b), 29, 116, 1643
392 A *šḫ·t*-staff. 46
393 *šš·t*-⌜fruit⌝.[184] 96, 108
394 *Šꜥ·t*-loaves.[185] 87
395 The terror (*šꜥ·t*) inspired by king. 900
396 *Šnś*-bread. 64
397 Libation (*ḳbḥ*). 22–23
398 *Kmḥ*-loaves.[123] 77
399 Bread (*tꜣ*). 61, ⌜64W⌝
400 The king's bread. 63(F 200), 217
401 Baked bread (*tꜣ ꜣšr*).[186] 78
402 ⌜—⌝ bread (*tꜣ ytḥ*). 60, 73
403 ⌜Fresh⌝ bread (*tꜣ wꜣḏ*). 107
404 *Nbś*-bread (*tꜣ-nbś*).[187] 99
405 ⌜—⌝ bread (*tꜣ syf*). 86
406 *Twꜣw·t*-oil. 51
407 The choicest of the (food)-offering (*tp(y·t)-wꜣḥ·t*).[147] 114
408 *Ttw*-bread. 60, 73
409 This ⌜bright red⌝ (*ṯmśś·t*) (crown ?), equated with king, who is then equated with the eye. 1460
410 ⌜—⌝ (*ṯr* or *ḥꜣṯr*). 66
411 Figs (*d(ꜣ)b*). 110
412 A *dp·t*-loaf. 38
413 The Red Crown (*dšr·t*). 901(F 17)
414 Red jars (*dšr·wt*). 249

[182] Cf. Budge (*Liturgy*, 67, 158, 211), following Maspero and Dümichen. The former prefers "butter" (Coptic

ⲥⲁⲓⲣⲉ

in Gen. 18:8), the latter "cheese." See Dümichen's discussion (*Patuamenap*, I, 19–20).

[183] Budge (*Liturgy*, 82–84, etc.) renders "breast."

[184] But Bollacher takes *šš·t* (later *šḫ·t*) for grain, and Maspero thinks it nuts.

[185] Distinction between two words *šꜥ·t*, one standing for cakes and the other for loaves, is remarked by Erman, *Zur äg. Wortforschung* (I), p. 15.

[186] On reading the *t* of this and the following references in its original value of *tꜣ*, "bread," cf. the variants of *tꜣ syf* and *tꜣ-nbś* given by Dümichen in *Patuamenap*, I, Pl. XXIV, No. 86, and Pl. XXVI, No. 112.

[187] Cf. Kamal in *Ann. du Serv.*, XII, 240.

F. Eye of Horus—

415 Adornment (*ḏbꜣ*), 1 ⸢*bšyꜣw*⸣-kilt.[188] 41

416 Milk (*ḏšr·t*).[189] 61, 88

417 ⸢Theban milk⸣ (⸢*ḏšr·t-Wꜣś·t*⸣). 89

418 Food: bread and beer(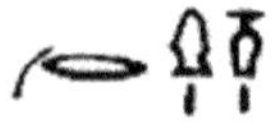). 60, 72

419 A barque. 1227(F 97)

420 Egypt or some part thereof. 1588–96(F 91, 57, 92, 279, 291, 85; C 94)

421 The king's head. 830(F 125)

422 The king himself. 698 etc. (F 130)

423 Cf. also the mystic names in F 18–26.

The eye of Horus is distinctly equated with the items of the preceding list, which therefore omits many objects mentioned in the funerary ritual.[190] The list may be summarized as follows:

Food: Breads, cakes, grains, meats, fowl, ⸢butter or cheese⸣, fruits.
Drink: Wines, beers, water, ⸢milk⸣.
Unguents.
Cosmetics.
Clothing.
Insignia: Crowns, scepters, staves.
Incense.
Libations.
Natron.
General terms for offerings.
Attributes or possessions of the king:
- His head.
- His soulhood and control.
- His terror (that inspired by him).
- His pyramid and mortuary temple.
- The king himself.

Miscellaneous:
- Olive-tree in Heliopolis.
- Willow-tree.
- Altar.
- *Wꜣg*-feast.
- Barque.
- Whole or part of Egypt itself.
- *Hꜣṯr* or *ṯr*.

VIII. Two Eyes of Horus

424 King is to take (*m*) the two eyes of Horus, the black and the white, to illumine his face. 33

425 Osiris the king is to take the two eyes of this great one and be satisfied with them (*ḥtp ḥr·śn*). 103

[188] Cf. label of such an object on coffin of Mentuhotep (Steindorff, *Grabfunde des Mittleren Reichs*, Pl. III).

[189] Sethe's rendering, quoted by Breasted.

[190] E.g., those identified with "the *ḥnḳ* which came forth from Osiris" (Pyr. 37, 39, 90, etc.).

F. Eye of Horus—

426 Geb has caused that Horus give his two eyes to Osiris the king, that latter may be satisfied with them. 583

427 ⸢King's son⸣ as Horus has come bringing (*yny*) the two bodily eyes of Horus to king to lead latter [to firmament unto Horus, to sky unto the] great [god, and to avenge] him on all his enemies. 69–70

428 [⸢King's son⸣ as Horus brings to king the two eyes of] Horus, which rejoice latter's heart (*pḏ·t(y) yb·f*). 71

429 King, after punishing his enemies with one of the eyes of Horus, returns it and causes that Horus see with both his eyes. 1240

430 Cf. also 96(F 166, 184), 108(F 166, 116), 253(E 11)

IX. UNRESTORED FRAGMENTS ALLUDING TO THE EYE

431 Eye mentioned in broken passages. 1923, 2166

G. Other Mythological References to Horus—

1 King has looked to Horus, Set, *Yꜣrw*, and the two *tt·ty-yb* as Horus looked to (*mꜣꜣ n*) Isis, as *Nḫb-kꜣ·w* looked to Selket, as Sobk looked to Neit, as Set looked to the two *tt·ty-yb*.[191] 487–89

2 Isis found her Horus and conducted his arm to Re to horizon. 741

3 Isis the Great bound on the girdle (*ṯs·t mḏḥ*, scil. upon her son) in Khemmis, bringing her ⸢censer⸣ (*dꜣyš*) and burning incense (*ydy*) before her son Horus the young child (*ẖrd nḫn*), that he might journey over the land in his two white sandals, going to see his father Osiris. 1214–15

4 Horus and his mother Isis each brought (*yny*) the other's heart (*yb*). 1640

5 Isis. See also 1199(G 17)

6 The cow that crosses the lake (*nm·t* ⸢*šꜣ*⸣) leads king to great seat born of the gods, born of Horus, begotten of Thoth (*yš·t wr·t yr·t-nṯr·w yr·t-Ḥr wtṯ·t-Ḏḥwty*). 1153

7 Horus opened mouth of Osiris with the adze of metal . . . 13–14(E 86)

8 Mouth of Osiris was opened (*wp*) by his beloved son Horus. 179

9 King's mouth is opened by Horus with his little finger (*ḏbꜥ nḏš*) wherewith [he] open[ed] the mouth of his father Osiris. 1330

10 Libations (*ḳbḥ·w*) of Osiris (here the king) came forth (*pr*) from his son Horus. 22

11 Osiris was caused to *šḏb* and to live by his various relatives: Atum, Shu, Tefnut, Geb, Nut, Isis, Nephthys, and Horus, as well as by the great and the little divine enneads as a whole; but as for his brothers Set and Thoth, he was caused to *šḏb* and to live and to punish (*ss*) them. 167–78

12 These (the text preceding) are the two charms (*ṯs·wy*) of Elephantine which were in mouth of Osiris, which Horus uttered (*ṯs*) concerning the vertebra (*bḳšw*). 234

13 Osiris ferried over to the firmament on a pair of reed floats (*sḫn·w*), with his son Horus beside him (*yr ḏbꜥ·wy·fy*, lit. 'at his two fingers') on another pair to sustain (*šnḫ*) him and cause him to dawn (*rdy ḫꜥ·f*) as great god in the firmament. 464–65

[191] In the cases of Horus, *Nḫb-kꜣ·w*, and Sobk at least, the looking is to the mother, apparently for sustenance and help.

G. Other Mythological References to Horus—

14 Horus avenged (*nḏ*) his father Osiris. 573(C 101), 758, 898(C 102), 1406(D 204)

15 Horus avenged his father. 633 and 1637(B 8), 1685(E 74), 21[91](D 27)

16 Horus made (*yry*) a *wḏꜣ·t*-garment for his father Osiris. 740

17 Osiris commended (*wḏ*) Horus to Isis on that day when he made her pregnant (*s·ywr*). 1199

18 Going forth (*pr·t*) of Osiris the king from his house is the going forth of Horus in embrace (*m sḫnw*) of Osiris the king. 1539

19 King's going (*šm·t*) is like the going of Horus to his father Osiris, that thereby latter might become a spirit and a soul, receive obeisance, and gain control. 1730

20 King's goings are the goings of Horus in embrace (*m sḫnw*) of his father Osiris.[192] 1860

21 Horus performed a crossing of the lake (*nm·t-⸢šꜣ⸣*)[193] for his father Osiris the king. 1550

22 King's ⸢equipment⸣ (*špd*) is upon him (*tp·f*), which Horus gave to Osiris. 1559

23 ⸢Priest⸣ (1st per.) brings (*yny*) to and puts on ((*w*)*d*(*y*) *n*) king the green cosmetic (*wꜣḏw*) which Horus put on his father Osiris. 1681–82

24 Horus embraced (*sḫn*) his father Osiris, (when) he found him upon his side in *Gḥsty*. 1799

25 Horus gave life (*rdy·f ꜥnḫ*) to his father, [presented] satisfaction ([*ꜣw·f*] *wꜣš*) to Osiris, before (*m ḫnt*) the Western gods. 1980

26 Osiris. See also 742(F 39), 956(E 28), 970(E 162), 971(D 251), 974–75(D 252), 1215(G 3), 1558(E 24), 1824(E 105), 2115(E 163)

27 The *šꜣw* of the Southern Palace (*pr wr*) runs for the Great One (Horus, cf. 853b); the Northern Palace flees before him (*rw n·k pr nsr*). 852

28 Earth is bidden to hear what Geb said (when) he was glorifying Osiris (here the king, cf. 1012a) as a god, (when) the watchers (*wrš·w*) of Pe made offering (*wdn*) to him and the watchers of Hieraconpolis honored (*sꜥḥ*) him like (*yś*) Sokar, ⸢Horus, *Ḥꜣ*, and *Ḥmn*⸣. 1013

29 Watchers. See also 795(G 39)

30 The Two Truths (*mꜣꜥ·ty*). 317(G 54)

31 Horus bound (*mr*) himself to his offspring (*msyy·t* etc.). 647

32 Morning-star, Horus of Dewat etc. is entreated to give to king his two fingers which he gave to the Beautiful One (*nfr·t*), daughter of the great god (*nṯr ꜥꜣ*) at separating (*wp·t*) of sky from earth, even at ascending (*pr·t*) of the gods to the sky, he being a soul (*bꜣ*) and dawning in the bow of his ship (described under D 282). 1208–9

33 *Nḫy*-serpent. 663(D 292)

34 The great god (*nṯr ꜥꜣ*), some solar deity. 1208(G 32)

35 Eastern gods (*nṯr·w yꜣbty·w*). 1209(D 282)

36 Western gods (*nṯr·w ymnty·w*). 1980(G 25)

37 Gods of Pe. 1209(D 282)

[192] Cf. with the foregoing 768, etc. (D 8–9).

[193] Cf. 1153(G 6). Ferryman is called *nm ⸢šꜣ⸣* in 1224.

G. Other Mythological References to Horus—

38 The gods in general. 795(G 39), 957(G 44), 1217(E 91), 1658(G 57)

39 Earth is bidden to hear what the gods said and Re said (when) he was glorifying king that latter might receive his spirithood before the gods like (*yš*) Horus, son of Osiris, when he (Re) was giving him (Horus) his spirithood which is among the watchers of Pe and was honoring him as a god among the watchers of Hieraconpolis (N repeats 'Pe' instead). 795

40 Horus expelled (*ḫšr*) Re from the south (*rš·t*) of the sky. 2158

41 Nile (*Ḥʿpy*). 2047(C 79)

42 Horus crushed (*ṯbb*) with his foot (*ṯbw*) the mouth of the *ḥfꜣw*-serpent, trapper of the phallus (*ḥb-ḥnn*). 681

43 Orion (*Sꜣḥ*). 261(E 21)

44 Set is bidden to remember and take to heart (*(w)d(y) yr yb·k*) the speech of Geb and the threat (*fꜣw*) which the gods made against him in the Prince-house in Heliopolis because of his felling (*ndy*) Osiris to earth, when Set said: "I did not do this against him," that he might prevail (*šḫm*) thereby over (*n* for *m*?) Horus. 957–58

45 Horus took (*yṯy*) estate (*pr*) of his father from his father's brother Set before (*m bꜣḥ*) Geb. 1219

46 The first body (*ḫ·t*) of the ⸢company⸣ (*ydr·wt*) of the triumphant (*mꜣʿ-ḫrw*) was born before the attack (*ḏnd*) took place, before the voice arose, before the reviling (*šnṯ·t*) occurred, before the conflict (*ḫnnw*) began (verb in each of these four phrases is *ḫpr*), before the eye of Horus was plucked out (*ḳny*), before the testicles of Set were wrenched off (*šꜣd*). 1462–63

47 Set. See also 850(D 160), 971(D 251)

48 Shu. 317(G 54)

49 *Šnṯ*-serpent. 681(C 80)

50 King is Horus who came forth from acacia (*šnḏ*) upon being commanded: "Guard thyself, O lion (*sꜣw ṯw rw*)!"[65] 436

51 Horus followed and loved Geb, while latter dawned as King of Upper and Lower Egypt controlling all the gods and their kas. 1625–26

52 Geb. See also 317(G 54), 957(G 44), 1219(G 45)

53 The Two Lands (*tꜣ·wy*) bowed before Horus and feared before Set. 57 (F 265)

54 King as Horus, the heir of his father, triumphs (*mꜣʿ ḫrw·f*) through that which he did. *Tfn* and *Tfn·t* have judged (*wḏʿ*) him; the Two Truths have held the (legal) hearing, with Shu for a witness, and have commanded that there revert to (*phr n*) him the thrones (*nš·wt*) of Geb. 316–17

55 The great cavern (*ṯpḥ·t wr·t*) of Heliopolis was opened (*wn*) for Horus of Letopolis (*ḫnty Ḫm*). 810

56 *Db·t-nḥḥ·wt*. 522(D 235)

57 Thoth put (*(w)d(y)*) the gods beneath Horus ... in the audience-hall (*ḏꜣd(w)*) and the ⸢court of justice⸣ (⸢*(w)d(y)-mꜣʿ*⸣). 1658

58 ⸢Horus made a sitting⸣ — — —, that the judgment might take place (*ḫpr wḏʿ-mdw*). 2088

59 Miscellaneous. See also 378(D 156), 551(D 289), 854(D 55), 905(E 90), 2094(E 174)

H. Miscellaneous—

I. USED IN TITULARY

1 "Horus" a designation used in royal titulary. [6], 7a, 8, 786
2 "Harakhte" thus used. 7b
3 "*Ḥr ⸢šḫm m⸣ Nbty*" thus used. 6–8 and 786(A 85)

II. DOUBLE HORUS[43]

4 Two green Horuses (*Ḥr·wy wꜣḏ·wy*), in obscure connection. 457
5 Double Horus (*Ḥr·wy*), presider over estates (*ḫnty·wy pr·w*), lord of food (*nb-ḏfꜣ·w*), great one (dwelling) in Heliopolis (*wr (y)m(y) Ywnw*), is asked to give bread and beer to king and to make flourish (*s·wꜣḏ*) the king, his offering-table (*wdḥw*), and his butchering-block. 695–96

III. FEASTS, ETC.

6 King celebrates his year(ly feast)s (*rnp·wt*)[194] ⸢in Field of Rushes⸣ like (*my*) Horus, son of Atum. 874
7 King is to rejoice (*ḥꜣby*) at every feast (*ḥb*) of Horus. 1672
8 Seasons (*y·tr·w*) of Harakhte. 1693(C 57)

IV. BROKEN PASSAGES UNRESTORED

9 ——— like (*my*) Harakhte.[195] 1384
10 ——— life for Horus. 1825
11 Horus has put for himself ———. 1857

[194] That *yry rnp·wt* here means "celebrate yearly feasts" rather than "spend years" is shown by comparison with 1950, where *yr·n·(y) rnp·wt·k* appears. This again is paralleled by *yr·n·(y) tp(y)·w-ybd·w·k* (*šmd·wt·k* resp.) in 657.

[195] This phrase does not appear in 347, whence part of the context is restored.

SUPPLEMENT

Offspring of Horus

NAMES

* Not expressly called "children" or "offspring" of Horus.

Ḥ3py, Dw3-mw·t·f, Ymśty, and *Ḳbḥ-śn·w·f.* *149, *552N, *1092, *1097, *1228N, 1333, 1339, *2101

Ḥ3py, Dw3-mw·t·f, Ḳbḥ-śn·w·f, and *Ymśty.* *552TM

Ḥ3py, Ymśty, Dw3-mw·t·f, and *Ḳbḥ-śn·w·f.* *1228PM, 1548, *1983

Ymśty, Ḥ3py, Dw3-mw·t·f, and *Ḳbḥ-śn·w·f.* *601, *1483, 2078

OCCURRENCES

See C 23–26,[196] also individual names in Appendix.

CLASSIFIED REFERENCES

A. Epithets—

1 *fdw ypw y3ḫ·w,* "these four spirits." 1092(*C 8*)

2 *fdw ypw nṯr·w mś·w-Ḥr mr·f,* "these four gods, the children of Horus, his beloved." 1548(*E 5*)

3 *fdw ypw rḫ·w ny-św·t ... mśw·t-Ḥr Ḫm(y),* "these four royal intimates, ... offspring of Horus of Letopolis." 2078

B. Magical or Mystic Name—

1 [*t(w)t·w,* "the assembled ones."] 1983

C. Relationships—

I. Genealogical

1 Filial relation to Horus is summarized in C 23–26.

2 Are grandchildren of king. 1983(*E 7*)

II. Position Occupied in Relation to Other Divinities

3 Are souls (*b3·w*) of Horus. 2101(D 208)

4 Are king's right side, which is in Horus; while *Ḥ(w)-ḏndrw, Ḫnty-w3ḏ·wy·fy,* Nephthys, and *Ḫnty-ʿn-yr·ty*ʾ are king's left side, which is in Set. 601

5 Are with king as he is ferried over to Field of Rushes. 1092(*C 8*)

6 Are royal intimates. 2078(*A 3*)

III. Relations on Part of Other Divinities

a) Actions

7 Horus uses his children to aid king. 24(E 137), 619(E 101), 637(E 102, *E 10*), 766(E 137), [1828](E 138), [1829](E 102, *E 10*)

8 *Ḥr·f-ḥ3·f* is bidden to ferry these four spirits over to Field of Rushes as comrades of king, two on each side. 1092

[196] This and similar citations in Roman type refer to the treatment of Horus proper; italics indicate citations from this Supplement.

C. Relationships—

b) Position Occupied

9 King is called their father. 1333(*E 3*), 1340(*E 11*)
10 King controls them (*shm ym·sn*). 1828(E 138)
11 King has joined himself with them. 647(C 24)
12 King is identified with each of them individually. 1097
13 King is one of these four gods. 1483

c) Attitude

14 Horus loves them. 1548(*E 5*)

D. Nature—

I. ATTRIBUTES AND POWERS

1 Are gods. 1483(*C 13*), 1548(*E 5*)
2 Are spirits (*yꜣḫ·w*). 1092(*C 8*)
3 Live on truth (*mꜣꜥ·t*). 1483

II. HABITAT

4 Are bidden to bring (for king's ascent) the barque "Eye of Khnum" which is in the Lily Lake (*mr n ḫꜣ* in P; but MN have *mr pw n ḥtm*, 'this lake of the *ḥtm*-bird'). 1228
5 Are ferried over with king to Field of Rushes (*sḫ·t-yꜣr·w*). 1092(*C 8*)
6 Guard land of Upper Egypt (*tꜣ šmꜥ*). 1483

III. ATTITUDE TOWARD OTHER DIVINITIES

7 Have loved king. 647

IV. PARTS OF BODY

8 *ḏbꜥ·w*, fingers. 1983(*E 7*)

V. EQUIPMENT

9 Lean (*twꜣ*) upon their *ḏꜥm*-staves. 1483[197]

E. Activities (All in Behalf of King)—

1 Count (*yp*) for him. 24 and 766(E 137)
2 Have smitten his enemy and made red his blow, have punished him and suppressed him of the evil odor (the enemy). 643
3 Are to protect life of (*stp sꜣ-ꜥnḫ ḥr*) their father Osiris the king from him who would cause that he suffer (*sḏb*) in presence of the gods. 1333
4 Are to smite Set and avenge Osiris the king on him. 1334
5 That which is in slayer of the king in the hinder parts is for these four gods, the children of Horus, his beloved. 1548
6 Bear (*wṯs*) king. 619(E 101)
7 King's assembled grandchildren have raised (*ṯs*) him, [washed his face], ⌜checked⌝ (*yꜥḥ*) his weeping, and opened his mouth with their wondrous (*byꜣ·w*) fingers. 1983

197 Cf. 339 (note 97).

E. Activities (All in Behalf of King)—

8 Expel hunger from belly and thirst from lips of king. 552
9 Bring to king his name of "Imperishable (*y·ḫm-śk*)." 2102(D 208)
10 Betake themselves under king, without any of them escaping, and carry (*f*ꜣ) him (in ascent?). 637, 1338, 1823, 1829
11 Are to carry and lead (*śšm*) their father Osiris the king (in ascent?). 1340
12 *Ḥꜣpy* and *Dwꜣ-mw·t·f* are the king's arms, enabling him to ascend at will to the sky; and *Ymśty* and *Ḳbḥ-śn·w·f* are his legs, enabling him to descend at will to the underworld (*nn·t*). 149
13 Bring to king the barque "Eye of Khnum" (for ascent). 1228(*D 4*)
14 Rise before (*ꜥḥꜥ n*) king and bind for him a rope-ladder whereon he ascends to Khepter in east side of sky. 2078–79
15 Are to recite "chapter of the king (*rꜣ n yt(y)*)" for him. 1334

F. Mythological—

1 Cf. references under E above, in most of which the king is clearly identified with Osiris.
2 Possibly cf. 258d: "The gods of East and West are satisfied with the great thing which came to pass in the embrace of the offspring of the god (*mśw·t-nṯr*)." This may refer to the *mśw·t-Ḥr*, for epithet *nṯr* stands independently for Horus (A 41).

APPENDIX

INDEX OF OCCURRENCES OF DIVINE NAMES IN THE PYRAMID TEXTS

Yꜣmw·t: 131

Yꜣḥš (=*Rꜣḥš* and *Yḥš*): 994

Yʿḥ: 732, 1001, 1104

Yw·š-ʿꜣ·š (Saosis): 1210

Ywn-mw·t·f: 1593(F 291), 1603 Cf. 571

Ymn (Amon): ⸢1540⸣ Cf. 446, 1095, 1712N

Ymśty: 149(D 180, *E 12*), 552(*E 8*), 601(*C 4*), 1092(*C 8*), 1097(*C 12*), 1228(*D 4*), 1333(*E 3*), 1339(*E 11*), 1483(*C 13, D 3, D 9, D 6*), 1548(*E 5*, D 136), 1983(*E 7, B 1*), 2078(*A 3, E 14*), 2101(D 208)

Ynpw (Anubis): 57, 135, 157, 220, 468, 574 (cf. C 101), 592, 727, 745, 793, 796–97, 804, 806–08, 896–97, 1014, 1015(C 85–86), 1019, 1122, 1162, 1180, 1257, 1282, 1287, 1295(D 88), 1335(E 59), 1364, 1380, 1523, 1537, 1549, 1552, 1564, 1672, 1676, 1713, 1723, 1833, 1867, 1995, 2001 (cf. D 50), 2012, 2026, 2069, 2150, 2178, [2184], 2198

Yḥš (=*Yꜣḥš* and *Rꜣḥš*): 1476

Yḫ·t: 179, 188–91, 198(C 1), 1147, ⸢1992⸣, ⸢2030⸣

Yḫ·t-wt·t: 198(C 1), 791, 1503, 1729, 2068, 2110

Yś·t[46] (Isis): 3, 32, ⸢123⸣, 155(E 121), 164, 172(G 11), 205, 210, 371, 379, 489(G 1), 556, 577, 584(C 52, E 62), 592, 606, 610, 628, 630, 632(C19), 707, 734, 741(G 2), 744, 755, 872, 898(C 102), 939, 960, 996, 1004, 1089, 1140 (C 96), 1154, 1199(G 17), 1214(E 3, G 3), 1255 (cf. C 51), 1263, 1265(E 180), 1272(E 181), ⸢1278⸣, 1280–81, 1292, 1302(D 151), 1347, 1358, 1362, 1375, 1464, 1472, [1521], 1547, 1630, 1634, 1635(C 19), 1640(G 4), 1655, 1703, 1750, 1873, 1886, 1951, 1964, 1973, 1981, 1997, 2009, 2089, 2098, 2144, 2188, 2192 Cf. F 144

Yśyr[46] (Osiris, name used independently): 13(E 86), 16, 17(D 191), 22(G 10), 29 (king's name omitted by error), 31, 39, 57, 62, 63, 105, 117–18, 134, 144 (C 4, D 22), 145(C 59, E 178), 146(E 179), 155(E 121), 161, 167–78(G 11), 179(G 8), 186, 215, 218 and 222(D 78), 234(G 12), 251, 308 and 312(F 144), 313, 350 (n. 145), 364, 443, 464(G 13), 466(C 89), 467(D 224), 472(E 112), 493(E 4), 517, 520(D 201), 523, 549, 573(C 101), 574, 622, 625, 658, 684(D 14), 691, 722, 740(G 16), 742(F 39), 749, 752, 754, 757, 758(G 14), 759, 778, 788, 790, 793, 795(G 39), 819, 826, 832, 848, 865, 872, 882–84, 895(C 99), 898 (C 102), 899, 925, 956(E 28), 957(G 44), 960, 963, 964 and 966–68 (n. 57), 970(E 162), 971(D 251–52), 972, 984(D 48), 988(D 65), 1004–5 (cf. C 27), 1012, 1013(G 28), 1033, 1035, 1044, 1068, 1090, 1122(D 89), 1128(E 5), 1157, 1194–95, 1199(G 17), 1201–2, 1215(G 3), 1232, 1236, 1256, 1258–59, 1264(E 180), 1267(E 181), 1280, 1282, 1291, 1297–98, 1328, 1330(G 9, C 90), 1354, 1358, 1360, 1362, 1368, 1392–93, 1395, 1406(C 5), 1419, 1428, 1448, 1450(F 142), 1486–87, 1500, 1502, 1505(C 83), 1516, 1520–21, 1523, 1525, 1549 (n. 72), 1551, 1556, 1558(E 24), 1559(G 22), 1567, 1643(F 56), 1655,

1657, 1658(C 5), 1665–66, 1672, 1681–82(G 23), 1683, 1730(G 19), 1748, 1761, 1799(D 261), 1800, 1803, 1804(E 146), 1833, 1860(G 20), 1882, 1978 (C 54), 1980(C 5), 2000, 2007, 2010, 2016, 2021–22, 2031, 2051, 2054, 2055 (n. 117), 2068, 2076, 2092, 2097, 2103(C 103), 2105, 2108, 2111, 2115(E 163), 2144, 2201 Cf. 8, 1142, 1256

Yśyr (Osiris, in apposition with name of king[198]): 8, 12, 15, 18, 19, 21, 24, 25, 30, 35–37, [43–44], [46–48], 50, 51, 54, 55, 59–61, 64–68, 72–103, 106–7, 110, 111, 113–15, 117, [249], 575, 582–83, 586, 590, 609, 612, 618–19, 626, 634, 636, 640, 645, 648, 651, 737, 765, 775–78, 786, 825, 831, 847, 849, 939, 960, 966aN (name of king added by error), [1002], 1003, [1004], 1012, 1046, 1069, 1142 and 1256 (name of king added by error), 1258–59, 1262, 1333–35, 1337–43, 1379–80, 1382, [1383–84], 1385, [1386], 1523, 1531, 1539, 1544, 1550, 1607, 1609–10, 1615–16, 1619–20, 1622, [1623], 1683, 1685–86, 1696–1700, [1708], 1753–56, 1787–88, 1794–95, 1798, 1805–10, 1812–14, 1817, 1819, 1822, 1824, [1826], 1828, 1830–31, 1838–40, 1870, 1872, 1894, [1899], 1973, 1975, 1986, 2033

Ytm (Atum): 124, 135, 140, 145(C 59), 147, 148–49(D 180), 151–52, 154, 156, 158, 160, 167(G 11), [199], 207, 211(C 22), 212–13, 229, 241, 258(D 138), 261, 301(D 263), 304, 305(D 210), 380, 395, 425, 447, 479, 492, 518, 603–5, 701, 840, 843, 874(H 6), 879, 881(D 56), 942–43, 961(D 86), 992, 997, 1173, 1237, 1241, 1248, 1277, 1297–98, 1302(D 151), 1347, 1376, 1447, 1451, 1466, 1473, 1475(D 87), 1489(D 25), 1521, 1525, [1546], 1587, 1617, 1645–47, 1652–56, 1660, 1686(E 145), 1694–95, 1742, 1778, 1818, 1870–71, 1982, 1984, 2009, 2024, 2057, 2065, 2081–82, 2098, 2146, 2163

ꜥnḏty: 220, 614, 1833 Cf. 182

Wꜣḏ·t (Buto): 702, ⸢792⸣, 1671, 1875 (n. 50)

Wp-wꜣ·wt: 126, 455, 463, 769, 953(D 232), 1009, 1011, 1066, 1090, 1287, 1304, 1374, 1379, 1438, 1638, 1979(E 173), 2032(C 98) Cf. 57, 643, 1215, 1239, 1806[199]

Wng: 607, 952(C 110)

Wnṯ (error for *Swnṯw*): 1094M

Wśr·t: ⸢270⸣

Bꜣby: 419, 502, 515–16, 1310, 1349 (n. 84)

Bꜣśty·t (Bast): 892, 1111, 1310

Ptḥ (Ptah): 560 and 566(D 207), 1482

Ftk-tꜣ: 120, 123, 545

Mꜣ-ḥꜣ·f: 383, 597, 599, 925, 1222, 1227(F 97), 1769

Mꜣꜥ·t: 319, 323, 1580, 1768, 1774 Cf. 265, 1306, 1429, 1483, 1775, also duals in 317, 1315, 1785

Mꜣfd·t: 230, 438, 440, 442, 677, 685, 1212(D 255)

Mw·t: ⸢123⸣, ⸢734⸣

Mnw (Min): 256, 424, 953(D 232), 1712 (cf. 1013), 1928(D 76), 1948, 1993 (C 105), 1998

Mnṯw: ⸢724⸣, 1081, 1378

Mḫnty-⸢yr·ty⸣ (=*Ḫnty-⸢yr·ty⸣*): 601, 826, 1265(E 180), 1431, 1864

[198] For citations in thesis, see Horus-treatment *passim*, especially §§ E and F.

[199] Wolf erect on standard is found also as determinative of *wpyw* in 1913, of *nwꜣ* in 13, of *sꜣb* in 804 and 1015, and of *šmś·w-Ḥr* in 921 and 1245.

*Mḫnty-ʿn-yr·ty*ʾ (= *Ḫnty-ʿyr·ty*ʾ): 771
N(y)·t (Neit, orig. *Nr·t*): 489(G 1), 510, 606, 1314, 1375, 1521, 1547
Nyw: 207, 446 Cf. *nyw*, "ostrich," in 469
Nw·t: 1–5, 7, 8, 25, 171(G 11), 208, 250(C 72), 275, 299, 383, 441, 459, 519, 530, 541, 543, 580, 597, 603, 616, 623, 626, 638, 698, 741, 756, 765, 777–80, 786, 802(E 108), 823(C 56 and n. 91), 824–25, 827, 834, 838, 842, 883, 902, 933, 941, 990, 1016, 1021, 1030, 1036, 1048 (n. 77), 1049, 1082, 1090, 1101, 1145, 1149, 1169, 1173–74, 1184, 1188, 1213(C 8), 1247, 1291, 1300, 1311, 1321, 1328, 1332, 1341, 1344, 1361, 1405, 1416–17, 1422, 1426, 1428, 1430, 1454, 1471, 1479, 1516, 1521, 1546, 1596(n. 167), 1607, 1629, 1655, 1664, 1688, 1702–3, 1720, 1758, [1833], 1835, 1895, 1960, 2028, 2034, 2037(D 62), 2041, 2052, 2057, 2091, 2107, 2150, 2171, 2178, [2206]
Nwnw (Nun): 132, 237, 268, 310, 314, 318–19, 392, ʿ426–27ʾ, 551, 593, 603–4, 606, 871–72, 1034, 1040, 1057, 1078, 1166, 1174, 1304, 1446, ʿ1460ʾ(n. 84), 1486, 1517, 1525, 1678, 1691, 1701, 1778, 1780, 1964, 2037, 2147
Nb·t-ḥt·t (Nephthys): 3, 150, 153(E 121), 164, 174(G 11), 203, 210, 371, 379, ʿ444ʾ(E 7), 556, 577, 584(E 62, C 52), 601(*C 4*), 606, 610, 616, 623, 628, 630, 707, 755, 872, 898(C 102), 939, 960, 996, 1004, 1089, 1154, 1255(cf. C 51), 1265(E 180), 1273(E 181), 1278, 1280–82, 1292, 1347, 1354, 1362–63, 1375, 1427, 1464, 1547, 1630, 1634, 1655, 1750, 1786, 1873, 1951, 1973, [1981], [1997], 2009, 2098, 2144, 2192 Cf. F 144
Nfr-tm: 266, 483
Nny and *Nnw*: 310, 314, 445, ʿ670ʾ Cf. verbs in 138, 163, 428, 541, 1595–96, 1605
ʿ*Nn·t*ʾ: 207, 446, 1691
Nḫb-kꜣ·w: 229, 340(n. 97), 346(C 35), 356, 361, 489(G 1), 1146, 1708 Cf. 161, 311, 315, 512, 2040
Nḫb·t: 696, 1229, 1451 Cf. 4, 123
Rꜣḥś (= *Yꜣḥś* and *Yḥś*): 1476
Rʿ (Re): 6–8, 34, 37, 120–24, 128, 130, 132, 136, 145(C 59), 152, 154, 156, 158, 160, [199], 200 and 209(cf. D 28), 226, ʿ227ʾ, 231, 263, 266–69, 273–75, 285, 310, 313–14, 328(C 112), 336, 337(D 31, D 36), 340(n. 97), 346(C 35), 348 (C 58), 351(D 32, D 36), 356–57, 358(D 32, D 36), 362(D 45), 366–68, 370, 372(E 94), 390(E 111), 392, 442, 449(D 162, cf. E 41), 452(B 4), 460–61, 470, 472(E 112), 473(E 166), 482–83, 490, 495, 517, 519(E 95), 531, 534, 542–43, 546, 560 and 566(D 207), 585, 597, 598(D 82), 599, 602, 607, 621, 636, 673(D 249), 698(F 280), 702(C 113), 703, 706, 709–11, 713, 721, 726, 730, 732–33, 741(G 2), 743, 750, 756–57, 760, 762, 787, 792, 795(G 39), 800(C 82), 804(C 85), 812–13, 819, 855–56(D 223), 886–89, 891, 893, 906, 915, 918–19, 922–23, 926(D 32), 927(D 37), 932(D 32), 933(D 37), 948, 950, 951(C 118, cf. A 48), 952(C 110), 953(D 232), 955, 971(D 251), 974–75(D 252), 990, 999–1000(D 39), 1016(E 160), 1029, 1044–45, 1049(D 60), 1063, 1084–86 (D 34), 1087(C 127, cf. E 169 and D 266), 1091, 1103(D 33), 1107–8, 1141–42, 1167, 1169, 1171, 1178–80, 1204, 1206, 1231, 1238, 1244(E 170), 1246, 1247(E 96), 1261, 1263, 1299, 1316–18, 1343, 1345, 1347, 1359, 1372, 1405, 1421, 1423, 1425, 1430(D 67), 1437, 1440, 1442, 1449, 1461, 1464, 1465 (E 118), 1469–70, 1471(n. 61), 1479, 1481, 1492–95(E 119), 1496–98, 1500, 1503, 1508(C 84), 1517, 1518(D 12), 1531–32, 1540, 1542, 1568, ʿ1572ʾ,

[1574], 1582, 1669, 1679, 1686(E 145), 1687–88, 1692, 1694–95, 1705, 1709, 1719(C 86), 1720, 1734, 1739, 1759, 1773–74, 1785, 1802, 1808, 1835, 1862 (D 200), 1863(C 104), 1877, [1887], 1906, 1991, ⸢2005⸣, 2019, 2025, 2028, 2035 (cf. D 29 and D 62), 2045 (cf. D 40), 2047(C 79), 2048, 2062, 2077, 2090(F 319), 2095, 2158(G 40), 2169, 2172, 2174, [2175], 2183, 2206, 2208, 2212[200] Cf. 137a, 1212, 1355 Cf. D 221

Rnn-wt·t: 302, 454(C 121) Cf. 1755, 1794

Ḥꜣ: 1013(G 28), 1712

Ḥꜣ·f-m-ḥꜣ·f: 517

Ḥꜣpy: same occurrences as *Ymšty*, *q.v.* Cf. also 279

Ḥꜣpy (Apis): 286, 1313, 1998

Ḥyḥy: 1390

Ḥmn: 235, 1013(G 28)

Ḥrw (Horus): 4–8, 10–14, 17–22, 24–29, 31–33, 35, 36, 38–40, 42, [43], 44, 45, [46–47], 48, 51, 52, 54–61, 63–74, 76–89, 91–101, 103–111, 113–18, 133, 135, 138–39, 141–46, 148, 159, 176, 179, 192, 195, 198, 206, 211, 216, 218, 222, 234, 244–45, [249], 250, 253, 257–58, 261, 295–96, 301, 304, 308, 312, 316, 330–31, 337, 342, 346, 348, 351, 353, 358, 360, 362, 372, 378, 390, 418, 436, 444, 449–50, 457, 465–67, 472–73, 480, 487, 489, 493, ⸢502⸣, 503, 518–20, 522, 525–28, 534–35, 551, 555, 560, 566, 573, 575, 577–87, 589–92, 594–96, 598, 600–601, 607, 609–15, 617–20, 632–37, 640–51, 653, 659, 663–64, 670, 673, 678–79, 681, 683–85, 695, 698, 702, 723, 734, 737, 740–42, 746, 758, 765–68, 770, 786, 793, 795, 798, 800–801, 804, 810, 815, 823, 830–31, 841, 844, 846, 850, 853, 855–56, 874, 877, 881, 888, 891, 895, 897–98, 900–901, 903, 905, 921, 926–28, 932–34, 943, 946–48, 951, 953, 956, 958, 961, 969–71, 973, 976–77, 981–83, 986–87, 994, 999, 1007, 1010–11, 1013, 1015–16, 1025–27, 1030, 1036, 1040, 1048–49, 1067–68, 1084–89, 1103, 1113, 1122, 1129, 1131–32, 1134, 1136, 1140, 1147–48, 1153, 1176, 1199, 1202, 1207, 1214, 1219, 1227, 1231–35, 1237, 1239–45, 1247, 1254, 1257–59, 1264, 1268, 1277, 1285, 1293–95, 1301–2, 1320, 1327, 1330–31, 1333–35, 1338–39, 1354–55, 1373, 1375, 1384, 1406–15, 1429–30, 1436, 1449–50, 1458, 1460, 1463, 1465, 1471, 1475, 1478, 1489, 1492, 1505, 1507–8, 1518, 1539, 1548, 1550, 1558–59, 1570, 1588–89, 1594, 1596, 1609, 1612, 1614, 1622, 1625, 1627, 1632–33, 1636–37, 1640, 1642–43, 1657–59, 1668, 1672, 1681–83, 1685, ⸢1686⸣, ⸢1690⸣, 1693, 1702, 1710, 1712, 1715, 1719, 1728, 1730, 1733–35, 1742, 1753–56, 1794–95, 1798–1800, 1803–8, 1813, 1823–25, [1826–28], 1831–32, 1838–40, 1843, 1857–58, 1860, 1863, 1881, 1915, 1923, 1928, 1945, 1951, 1959, [1976], 1978–80, 1988, 1993–94, 2011, 2019, 2032–33, 2036–37, 2046–47, 2050, 2056, 2062, 2071–72, 2074–76, 2078, 2087–91, 2094, 2099–2101, 2103, 2106, 2115, 2147, 2158, 2166, 2185, 2190–91, 2202, 2213[201] Cf. 917a, also *Ḥt·t-Ḥr*

Ḥr·f-ḥꜣ·f: 383, 999, 1091(*C 8*), 1201, 1227(F 97), 1441, ⸢1585⸣ Cf. 493

Ḥsꜣ·t: 1029, 2080

Ḥḳ·t: 1312

[200] *Rʿ* occurs also *passim* in the royal names *Mryy-Rʿ* (later throne-name of Pepi I), *Mr·n-Rʿ*, and *Nfr-kꜣ-Rʿ* (throne-name of Pepi II).

[201] *Ḥr* occurs also in the name *Nfr-sꜣ-Ḥr*, early throne-name of Pepi I, which stood originally in 868c, etc. See Sethe, *Pyramidentexte*, I, xii, and Möller in *Zeitschrift für äg. Sprache*, XLIV, 129.

Ḥkn-wt·t: 288
Ḥt·t-Ḥr (Hathor): 466(C 89), 546, 705, 1278 Cf. 1025–27(D 58), 1327(D 59)
Ḥḏ·t-wt·t: 900
Ḫꜥy-tꜣw: 242, 423, 518
Ḫprr: 199, 305(n. 110), 888, 918, 1210, 1445, ⸢1546⸣, 1587, 1652, 1695, 1757, 1874, 2079(*E 14*), 2083, [2206] Cf. noun "beetle" in 366, 561, 570, 697, 1301(D 3), 1633, 2107, also *dbḥ ḫpr* in 1771 and 1777
Ḫnsw: ⸢402⸣ Cf. verb in 130, 456, 748, 798, 881, 1049, ⸢1306⸣, 1510, 1984
Ḫnty-Ymnty·w: 57, 133(C 61), 139(F 161), 220, 474, 592, 650, 745, 759, 811, 818, 869, 1145, 1393, 1746, 1748, 1833, [1851], 1942, 1996, 1999, 2021, 2198 Cf. 1666
Ḫnty-⸢yr·ty⸣(=*Mḫnty-⸢yr·ty⸣* etc.): 17(D 191), 148(D 180), 771, 826, 832, 1211, 1270(E 181), 1367, 1547, 1670, 2015, 2086
Ḫnty-⸢n-yr·ty⸣ (=*Ḫnty-⸢yr·ty⸣*): 601(*C 4*), 771
H̱nmw (Khnum): 445, 524, 1227(F 96), 1228, 1238, 1769
H̱rty: 350(n. 145), 445, 545, 1264(E 180), 1308, 1547, 1557, 1905[202]
Swntw: 1019, 1094, 1152, 1250
Spꜣ: 17(D 191), 27–28(D 267), 244, 254, 425, 444, 663, 669, 1613(D 284) Cf. 1098, 1452, 1470, 2069
Skr (Sokar): 241, 445, 620, 990, 1013(G 28), 1256, 1289, 1356, 1429, 1712, [1824], 1826, 1968, 1998, 2042, 2069 Cf. B 2, B 5, C 97
Śꜣḥ (Orion): 151, 186, 261(E 21), 408, 723, 802(E 108), 819–21, 882–83, 925, 959, 1436, 1561, 1717, 2116, 2172, 2180 Cf. 1763
Śbk: 456(D 155, D 275, n. 67), 489(G 1), ⸢507⸣, 510, 1564
Śpdw: 148(D 180), ⸢270⸣, 480, 994, 1476, 1863(C 104) Cf. 201, 1159, 1534
Śpd·t (Sothis): 151, ⸢270⸣, 341, 357, 363, 458(n. 49), 459, 632(B 6, C 19), 723, 822, 929, 935, 965, 1082, 1123, 1152, 1428, 1437, 1482(n. 49), 1561, 1636 (B 6, C 19), 1707
Śrḳ·t (Selkis): 183, ⸢227⸣, ⸢234⸣, 489(G 1), 1061, 1273, 1314, 1375PM, 1427, 1435, 1469, 1547 Cf. *Śrḳ·t-⸢ḥtw⸣* in 606, 673(D 249), 1375N
Śḫm·t: 262(n. 83), 1547
Śšꜣw: 426
Śšꜣ·t: 616
Škšn: ⸢498⸣, 1440, 1734(C 93), ⸢2186⸣
Štš (Set): 14(E 86), 17(D 191), 20(F 288), 26(E 88), 27–28(D 267), 36 and 39(F 111), 48(F 115, F 276), 57(F 265), 61(F 110), 65(F 90), 73(F 117), 84(F 246), 88(F 110), 95(F 83), 128, 135(D 75), 141(A 6), 142(D 168, C 21), 144(C 4, D 22), 153(E 121), 163, 173(G 11), 204(D 23), 205(D 274), 211(C 22), 218 and 222(D 78), 261(E 49), 294, 390(E 111), 418(F 66), 473(E 166), 480(D 79), 487(D 81), 489(G 1), 518(D 129), 535(D 198), 575–76(E 56), 580, 581(E 50, cf. E 78), 587(E 73), 591(F 51), 592(E 70), 594(F 30, F 278), 595–96(F 100), 598(D 82), 601(*C 4*), 678(E 14), 679(F 67), 683(D 17), 685(C 77), 734(D 108), 746(E 97), 770(D 83), 777, 793(C 67), 798(D 9), 801, 823(C 56, n. 91), 826, 832, 850(D 160), 865, 915–16, 943 (D 84), 946(F 98), 948(D 85), 957–58(G 44), 959–60, 961 (D 86), 971(D 251),

[202] *H̱rty* appears later in proper names also, e.g., *Sꜣ-H̱rty* and *Sꜣ·t-H̱rty* on Brit. Mus. stela No. 224 and *H̱rty-m-sꜣ·f* on No. 308 (*Hierogl. Texts* . . . , II, Plates 28 and 29, XII. and XIII. Dyn. respectively, both from "Sams Collection, 1834").

972, 974–75(D 252), 979, 994(D 80), 1016(E 160), 1033, 1035, 1067, 1145, 1148(E 104, C 69), 1150, 1219(G 45), 1233(F 107), 1236, 1242(F 204), 1256, 1258, 1259(C 67), 1264(E 180), 1269(E 181), 1285(C 66), 1309, 1334(*E 4*), 1407(F 259), 1453, 1459, 1463(G 46), 1465(E 118), 1467, 1475(D 87), 1487, 1493(E 119), 1500, 1521, 1556, 1594(F 291), 1595(F 85), 1612(D 284), 1628, 1632(E 55), 1655, 1667, 1699, 1710(C 68), 1715(D 9), [1735](D 75), 1742(F 226), ⸢1756⸣(F 316), 1839(F 218), 1904, 1906, 1928 (D 76), 1993, 1999, 2038, 2047(C 79), 2071(D 258, F 35), 2099(D 77), 2100(E 175), 2162, [2213](F 107) Cf. A 85; F 106, 109, 113–14, 116, 179–80, 183–84; n. 65

Šty·t (Satis): 812, 1116

Šw: 5, ⸢123⸣, 125, 168(G 11), 208, 275, 288, 294, 299, 313, 317(G 54), 324–25, ⸢427⸣, 447, 519, 531, 552–53, 593, 603–4, 677, ⸢692⸣, 784, 842, 1022, 1039, 1066, 1090, 1101, 1121, 1151, 1247–48, 1353, 1421–22, 1425, 1430, 1443, 1454, 1471, 1521, 1546, 1553, 1569, 1615, 1652, 1654–55, 1661, 1691, 1739, 1817, 1870–72, 1953, 1985, 1992, 2053, 2065, 2091(F 119–20), 2099

Šsmw: 403, 545, 1552(n. 72)

Ššꜣ: 1329, 2080 Cf. 127, 2154

Ḳbḥ-šn·w·f: same occurrences as *Ymšty, q.v.*

Ḳbḥw·t: 687, ⸢792⸣, 1180, 1285, 1348, 1564, 1749, 1995, 2103

Gꜣswt(y): 2080

Gbb (Geb): 1–3, 7–9, 80, 101–2, 138, 139(F 121), 144(C 4, D 22), 162, 170(G 11), 218, 231, 255, 258, 277, 301(D 263), 308, 312, 316, 317(G 54), 324, 398, 439, 466(C 89), 477, 480(D 79), 483, 541, 576–78, 583(F 426), 590 and 612(C 62), 626, 634(C 62), 639, 640(C 63), 649, 655, 657, 675, 698, 779, 783, 787, 793, 796, 840, 843, 895(C 99), 942–43, 957(G 44), 961, 967, 973(D 234, C 91), 977 (F 59), 993, 1012, 1013(G 28), 1014, 1018, 1030, 1032–33, 1039, 1045, 1115, 1120, 1142, 1149, 1163, 1175, 1195, 1204, 1210, 1219(G 45), 1235, 1259, 1264, 1277–78, 1296, 1300, 1321, 1327, 1343, 1353(D 269), 1367, 1395, 1448, 1465, 1475(D 87), 1489(D 25), 1494(E 119), 1510, 1513, 1521, 1538, 1540, 1546, 1596, 1615, 1616(n. 61), 1620(cf. C 76, G 51), 1627, 1643(F 48), 1645, 1649, 1655, 1663, 1672, 1689, 1710, 1713, 1727, 1810–11, 1814, 1830, 1833–34, [1868], 1883, 1971, 1986, 1992, 2014, 2087(F 122), 2096, 2103(C 103), 2111, 2113, 2132, [2141], 2145, 2169 Cf. E 11 and n. 54

Tꜣy·t: 56(F 264), 738, 741 Cf. 737(F 149), ⸢816⸣, 1642(F 150), 1794 and 2074 (F 249)

Tby: 290, ⸢1394⸣

Tfn: 317(G 54)

Tfn·t: 5, 169(G 11), 288, 317(G 54), 447, 552–53, 779, 842, 990, 1066, 1248, 1353, 1405, 1443, 1521, 1546, 1652, 1654–55, 1662, 1691, 1739, 1985, 2053, 2065, 2099

Dwꜣw: 480, 994, 1155, ⸢1480⸣

Dwꜣ-wr: 1329, 1428, 2042

Dwꜣ-mw·t·f: same occurrences as *Ymšty, q.v.*

Dwꜣ-nṯr (Morning-star): 132, 357, 631, 732, 805, 929, 935, 1001, 1104, 1123, 1207(D 61; cf. G 32, D 282, D 255, D 181, E 3, E 91–92, E 167, E 142), 1295(D 88), 1366, 1372, 1707, 1719, 2005, 2014 Cf. *sbꜣ-dwꜣw* in 871

Ddy: 673(D 249, cf. E 14), ⸢1070⸣

Ddwn: 803, 994, 1017, 1476, 1718

Dnn-wt·t: 321

Drt: 486, 700

Dhwty (Thoth): 10, 16, 17(D 191), 27–28(D 267), [43](F 294), 58(F 124, 126, 128), 126, 128, 130, 157(E 121), 163, 175(G 11), 329(C 112), 387, 420, 468, 519(E 95), 535(D 198), 575(E 56), 594(F 278), 595–96(F 100), 635, 639, 651(E 57), 709, 796, 830(F 125), 956(E 28), 962, 976(F 258), 1089(D 285), 1153(G 6), 1176(D 175), 1233, 1235, 1237, 1247(E 96), 1254, 1265(E 180), 1271(E 181), 1305, 1336(E 59), 1377, 1429(E 117, D 119), 1465, 1507(D 287), 1523, 1570(D 152, cf. E 144), 1613(D 284), 1658(G 57), 1713, 1725, 1979 (E 60), 1999, 2118, 2150, 2213(F 127)

: 119

: 905P(E 90)

: 2080

: in ʿ—ʾ-*m-s³·f*, throne-name of Mernere, *passim*

www.ingramcontent.com/pod-product-compliance
Lightning Source LLC
Chambersburg PA
CBHW071228160426
43196CB00012B/2457
9781955087278